T0213509

Springer Biographies

The books published in the Springer Biographies tell of the life and work of scholars, innovators, and pioneers in all fields of learning and throughout the ages. Prominent scientists and philosophers will feature, but so too will lesser known personalities whose significant contributions deserve greater recognition and whose remarkable life stories will stir and motivate readers. Authored by historians and other academic writers, the volumes describe and analyse the main achievements of their subjects in manner accessible to nonspecialists, interweaving these with salient aspects of the protagonists' personal lives. Autobiographies and memoirs also fall into the scope of the series.

More information about this series at
https://link.springer.com/bookseries/13617

Markus Raffel · Bernd Lukasch

The Flying Man

Otto Lilienthal—History, Flights and Photographs

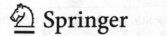

Markus Raffel 🆔
DLR Institut für Aerodynamik
and Strömungstechnik
Göttingen, Niedersachsen, Germany

Bernd Lukasch
Otto-Lilienthal-Museum
Anklam, Germany

This work contains media enhancements, which are displayed with a "play" icon. Material in the print book can be viewed on a mobile device by downloading the Springer Nature "More Media" app available in the major app stores. The media enhancements in the online version of the work can be accessed directly by authorized users.

ISSN 2365-0613   ISSN 2365-0621   (electronic)
Springer Biographies
ISBN 978-3-030-95035-4   ISBN 978-3-030-95033-0   (eBook)
https://doi.org/10.1007/978-3-030-95033-0

This Springer imprint is published by the registered company Springer Nature Switzerland AG
The registered company address is: Gewerbestrasse 11, 6330 Cham, Switzerland

Images that traveled around the world: man can fly! © Otto-Lilienthal-Museum.

Preface

Otto Lilienthal was the first person to make well-documented, repeated, successful flights with gliders and became known as the *flying man*. This book is the long-overdue first modern, English-language biography of the German pioneer. In some sections, it follows the great aeronautical biography of Otto Lilienthal, written by Werner Schwipps in 1988 under the title *Der Mensch fliegt* (*Man can fly*). In 1966, Schwipps had published his first biography of Lilienthal. Many others followed, dedicated to specific aspects of Lilienthals lifes work. Schwipps' books have remained standard references for research on Lilienthal to this day.

In 1991, a museum opened in Lilienthal's hometown of Anklam in northern Germany, in which all the flying machines of the aviation pioneer were reconstructed. Its long-time director is one of the authors. The detailed drawings of Lilienthal's flying machines are taken from the estate of Stephan Nitsch, held by the *Otto-Lilienthal-Museum*. After 2016, the occupation of European aviation research with Lilienthal's flight technology attracted worldwide attention. Markus Raffel, the second of the authors, is Professor of Aerodynamics at *Leibniz University of Hanover* and Head of the *Department of Helicopters* at the *Institute of Aerodynamics and Flow Technology* at DLR, the *German Aerospace Center*. After practicing hang gliding in France and California, he practically flew various authentic replicas of Lilienthals glider constructions.

Göttingen, Germany   Markus Raffel
Anklam, Germany   Bernd Lukasch

Acknowledgments

Many friends and aviation specialists made this book possible. In addition to Werner Schwipps, whose archives are now in the *Otto-Lilienthal-Museum* and the *German Museum of Technology* in Berlin, more recent research has been carried out especially at the *German Museum* in Munich and the *German Aerospace Center* (DLR). The authors would particularly like to thank Prof. Rolf Henke and Prof. Andreas Dillmann from the DLR, Charlotte Holzer from the *German Museum*, Tom Crouch from the *National Air and Space Museum*, Lewis Wyman from the Manuscript Division at the Library of Congress, Andrew Beem of *Winsports* in Los Angeles, the owner of *Kity Hawk Kites*, the hang gliding school on the Outer Banks, John Harris and his team of flight instructors lead by Billy Vaughn.

Special thanks go to Simine Short, author and aviation specialist at the *National Soaring Museum* in Elmira/New York, who has been in close contact with the authors for some time and has checked the manuscript critically. The authors would like to extend their sincere thanks to the editors Hal Brian (EAA), Angela Lahee (Springer), and Johannes Braukmann for their assistance at the final stage of manuscript preparation and for their insightful comments and suggestions.

The chapter *To fly is everything* contains excerpts from Paul Glenshaw's article *More Than a Century Later, Lilienthal and Wright Gliders Fly Together for the First Time at Kitty Hawk* originally appeared in *Air and Space/Smithonian Magazine*, January, 2020, and is used with his permission.

The poem of Otto Lilienthal, missing in the 1911 English edition, was translated by M. Buckow and G. Evans, Greifswald.

The annexes about the wind tunnel and flight tests of the patented monoplane, the large biplane and the experimental monoplane replicas are excerpts from articles published in the *AIAA-Journal of Aircraft* by Markus Raffel and his co-authors Felix Wienke, Pascal Weinhold, Clemens Schwarz, and Andreas Dillmann from DLR.

The reconstruction and detailed drawings of Lilienthal's flying machines are taken from the estate of Stephan Nitsch, held by the *Otto-Lilienthal-Museum*.

About This Book

"Of all the men who attacked the flying problem in the nineteenth century, Otto Lilienthal was easily the most important. His greatness appeared in every phase of the problem. No one equaled him in power to draw new recruits to the cause; no one equaled him in fullness and dearness of understanding of the principles of flight; no one did so much to convince the world of the advantages of curved wing surfaces; and no one did so much to transfer the problem of human flight to the open air where it belonged."

These words were spoken by Wilbur Wright, who successfully accomplished the first powered flight together with his brother Orville in 1903 on the sand dunes of the Outer Banks off the coast of North Carolina.

Lilienthal attracted worldwide attention due to a series of spectacular photographs showing him in flight, images made possible by technology that had only just been developed. This fortuitous union between a pioneer of aviation and the pioneers of so-called "instantaneous photography" is responsible for the immense contemporary popularity of Lilienthal's flights around the globe, the first ever free and successful heavier-than-air flights performed by man. This book traces the life of the German aviation pioneer, focusing on the designs of his many aircraft and the photographic documentation that has survived. The book also concludes with a spectacular research project conducted by one of the authors, right up to and including his own training exercises with exacting replicas of three of Lilienthal's designs. This project offered new insight into Lilienthal's work, but it also allowed for a spectacular meeting to unfold between replicas of Lilienthal's 1895 biplane and the Wright brothers' 1902 biplane in the air at a historic location on the Outer Banks in North Carolina.

Contents

About the Authors

Markus Raffel (*1962) is Professor of Aerodynamics at *Leibniz University of Hanover* and Head of the Department of Helicopters at the *Institute of Aerodynamics and Flow Technology* at DLR, the *German Aerospace Center*. He started at DLR in 1991, working in the field of experimental aerodynamics. Raffel is the recipient of several science awards from German, French, and US research organizations (e.g., *German Metrology Institute, French National Centre for Scientific Research, American Helicopter Society*). He has published more than 150 journal and conference papers and is the leading author of the widely distributed textbook *Particle Image Velocimetry*, about the most common optical flow diagnostic technique today.

After a short flying course with glider aircraft, he became interested in flying. He obtained his private pilot certificate and flies single engine and light-sport aircraft in Europe and in the USA. However, after practicing hang gliding in France and California, in preparation for flying Otto Lilienthal's monoplane, he admits that, for him, gliding with ultralight aircraft is the most exhilarating way of flying. He continues to enjoy flying whenever he finds time to do so.

Bernd Lukasch (*1954) studied physics at the *Humboldt University* of Berlin and earned his doctorate in 1984 after completing his thesis on *Atomic Collisions in Solid State Physics*. In 1988, he helped found the *Otto-Lilienthal-Museum* in Anklam which opened in 1991. Lukasch took over the management of the museum in 1992 and remained its director until 2019. Under his leadership, the museum was awarded the title *National Memorial* by the State Minister for Culture and Media of the Federal Republic of Germany and the *European Museum of the Year Award—Special Commendation* by the *European Museum Forum*.

Bernd Lukasch is the author of a biography about both Lilienthal brothers *Erfinderleben* (*inventors' lives*, 2005) and the author/editor of numerous publications in popular and history journals.

The Beginning of an Era

Why are we interested in the past? Why do we study history? If we could, we would undoubtedly try to look into the future also and study what's to come. But, sadly, we don't yet have this power. Historians look back because, as the saying goes, *"the future comes from the past"*. We don't know whether today's news, with so much noise and sensationalism, includes even one thing that might still be remembered in 10 years, much less 100. Something that would make people say, *"That's when it started. Nobody knew it at the time, but that's when the world changed."* For subsequent generations, certain names, dates, and events become synonymous with turning points, signposts of the end of one era and the beginning of the next—Archimedes, Galileo, Columbus, Einstein, and many more. The twentieth century has gone down in history as a century of world wars and great social upheaval, but also one of industrial and technological revolutions. Most importantly for our purposes, it was the century that brought us air and space travel.

The *"conquest of the heavens"* began more than 200 years ago with two French brothers named Montgolfier. In 1783, humanity defied gravity when a balloon of the brothers' design carried two people on the first successful manned flight in recorded history. The appeal of flight remains as strong today as it ever was—in Europe, they still call balloons *"Montgolfieres"* and celebrate November 21, the date of the first manned ascent, as *"Montgolfier Day"*. Hundreds of years earlier, the Chinese regularly flew man-carrying tethered kites, and largely apocryphal stories abound of those who affixed

M. Raffel and B. Lukasch, *The Flying Man*, Springer Biographies, https://doi.org/10.1007/978-3-030-95033-0_1

crudely built wings to their arms or backs and took leaps of faith that invariably ended badly.

We all know that date of the dawn of the century of air and space travel—December 17, 1903, when the Wright brothers made the first successful, documented, and controlled powered flight on the sand dunes of Kitty Hawk in North Carolina. But, when you dig deep, world-changing accomplishments are never as simple—nor as uncontested—as history eventually remembers them. People like Smithsonian Secretary Samuel Pierpont Langley in Washington, D.C., and the subsequently controversial Gustave Whitehead of Connecticut, not to mention George Cayley in England, Clement Ader in France, and Richard Pearse in New Zealand had experimented with what notable inventors like Alexander Graham Bell and Thomas Edison called "*the flying machine problem.*" But it was Orville and Wilbur Wright who made history, not because they flew 852 feet that cold December day, but because the age of aviation began with them. While the world was initially skeptical—news travelled a lot more slowly back then—once the Wrights began publicly demonstrating their aircraft in the United States and particularly in Europe, starting in France in 1908, their achievements were recognized, and Kitty Hawk and 1903 were enshrined in the history books. Powered and controlled human flight was proven possible, and the ancient myths and superstitions that had surrounded mankind's eternal dream to fly like a bird gave way to the technological history of the airplane.

The Wright brothers worked with remarkable efficiency, discipline, and determination. They analyzed, tested, and expanded on the world's body of aeronautical knowledge in less than a decade. They gleaned information wherever they could, but generally credit three key sources as they studied the nascent sciences of aviation: Octave Chanute of Chicago, editor of the *American Engineer and Railroad Journal*, and its spinoff called *Aeronautics*; Frenchman Louis-Pierre Mouillard, author of the book *The Empire of the Air*; and Germany's Otto Lilienthal who published the book *Birdflight as the Basis of Aviation* in 1889, and had himself been flying a series of pioneering gliders since 1891. The astonishing photos of Lilienthal made news around the world from 1893 on, and the Wright brothers themselves said that his fatal accident in 1896 was the catalyst they needed to turn their attention to flying.

While Lilienthal's book was not translated into English until 1911, the Wrights acquired a German copy, with a few key translated passages provided by Chanute. In 1912, Wilbur Wright wrote the following about Lilienthal:

"Of all the men who attacked the flying problem in the 19th century, Otto Lilienthal was easily the most important. His greatness appeared in every phase of the problem. No one equaled him in power to draw new recruits to the cause; no one equaled him in fullness and clearness of understanding of the principles of flight; no one did so much to convince the world of the advantages of curved wing surfaces; and no one did so much to transfer the problem of human flight to the open air where it belonged.

As a missionary he was wonderful. He presented the cause of human flight to his readers so earnestly, so attractively, and so convincingly that it was difficult for anyone to resist the temptation to make an attempt at it himself, even though his sober judgment and the misfortunes of all predecessors warned him to avoid touching it. If Lilienthal had done nothing more than this he still would have been one of the greatest contributors to the final success. But he was much more than a mere missionary.

As a scientific investigator none of his contemporaries was his equal. He set forth the advantages of arched wings in such convincing manner as to make him the real originator of this feature. [...]

Lilienthal was the real founder of out-of-door experimenting. It is true that attempts at gliding had been made hundreds of years before him, and that in the nineteenth century, Cayley, Spencer, Wenham, Mouillard, and many others were reported to have made feeble attempts to glide, but their failures were so complete that nothing of value resulted.

[... Whatever Lilienthal's] limitations may have been, he was without question the greatest of the precursors, and the world owes to him a great debt."

By the end of the nineteenth century, Lilienthal had successfully flown hundreds of times over distances of up to 800 feet in a series of more than a dozen gliders of his own design. In addition, he'd also manufactured and sold a series of gliders, one of which was purchased in 1896 by legendary publisher William Randolph Hearst for promotional purposes and survives to this day on display at the *Smithsonian Institution's National Air and Space Museum* in Washington, D.C.

Lilienthal was a successful designer and pilot, and had even been experimenting with different types of propulsion and control systems. Does this mean that the Wright brothers' fame is misplaced? In a word, no. Of course not. But it's hard to say how successful they would have been without the groundwork, and, perhaps more importantly, the inspiration provided by Lilienthal.

Lilienthal deciphered the secrets of lift and proved that the wings he designed could carry a person. He studied birds and learned to fly like one, bringing the ancient myth of Icarus to life. Lilienthal's studies and experimentation led him to make significant contributions to the young science of aerodynamics. But it wasn't science that made him famous—it was art. Ottomar Anschütz, Lilienthal's most prominent photographer, had, like his pioneering contemporaries, embraced the concept of short exposure photography, using new equipment he'd invented himself. These "snapshots" had supplanted the work of portrait painters over the years, with photographs quickly becoming the new standard for documentation. Anschütz inscribed the phrase *Nach dem Leben aufgenommen*, which translates to *captured true to life* on each photo he produced of one of Lilienthal's flights. The message was "man can fly", and it spread rapidly around the world, armed with photographic proof. It was a lucky coincidence that these developments in photography were happening at the same time as Lilienthal's pioneering flights. Without the photographs, it seems unlikely that his book would have ever piqued the Wright brothers' interest.

For those who saw them in person, Lilienthal's flights were an impressive spectacle, but they were seen more as an artistic achievement, as opposed to the dawn of a new technological age. Lilienthal crashed on August 9, 1896 and died the next day. The *flying man* was gone, and a vital chapter in the development of aviation had come to an end. Lilienthal's achievements were largely forgotten until the Wright brothers made their own flights in Germany more than a decade later. By then, Lilienthal's built-up flying hill, runway, and hangar had fallen into disrepair.

It is amazing to reflect on the fact that barely a quarter of a century passed between Lilienthal's gliding tests and the airborne battles of World War I. The technology of aviation has progressed at an almost unimaginable pace, while the idea of flight remains the dream of so many. Lilienthal's efforts, like aviation itself, represent both the romance of man's visceral, age-old yearning to fly, and the science that it took to make it a reality.

When Lilienthal imagined the future of flying, commercial flights and military aviation were not what he had in mind. To him, flying would be a sport (similar to the sport of soaring today), even a cultural institution that existed to bring those ancient dreams to life. He occasionally wrote about these predictions, imagining a future where the ability to fly was so profoundly world-changing that it broke down barriers between people and nations.

"*I, too, have made it a lifelong task of mine to add a cultural element to my work that should result in uniting countries and reconciling their people. Our experience of today's civilization suffers from the fact that it only happens on the surface of the earth. We have invented barricades between our countries, custom regulations and constraints, and complicated traffic laws. These are only possible because we are not in control of the 'kingdom of the air', and not as 'free as a bird'.*

Numerous technicians in every nation are doing their utmost to achieve the dream of free, unlimited flight and it is precisely here where changes can be made that would have a radical effect on our whole way of life. The borders between countries would lose their significance because they could not be closed off from each other. Linguistic differences would disappear, as human mobility increased. National defense would cease to devour the best resources of nations as it would become impossible in itself. And the necessity of resolving disagreements among nations in some other way than by bloody battles would, in its turn, lead us to eternal peace. We are getting closer to this goal. When we will reach it, I do not know."

Lilienthal's utopian vision of the benefits of aviation makes it all the more tragic that Germany's collective memory of him and his legacy is so closely tied to the two world wars, not to mention the many ways in which his story was coopted by the *National Socialists* for propaganda purposes. Lilienthal was idolized as a hero by the German air forces, which perhaps explains why no comprehensive biography of the man has been published in English to date, though there are many in German. Lilienthal's legacy and body of work have been studied, and his scientific and technical achievements have been scrutinized and published in biographical form—a selection of these writings is listed in the appendix. Besides the history of aviation itself, several books have also been written about the many-faceted personalities of Otto and his brother Gustav. Werner Schwipps, a scientific journalist from Cologne, has written about Lilienthal from multiple different perspectives for decades. In Germany, he is considered the father of Lilienthal biographers. The latest of his biographies was published in 1988, entitled *Der Mensch fliegt* or *Man Can Fly*—Historic Photographs of Lilienthal's Flight Experiments. Schwipps' book provides the foundations of this one—the first ever comprehensive English-language biography of Lilienthal. Since then, the previously incomplete photographic legacy of Otto Lilienthal has been extensively studied at the *Otto-Lilienthal-Museum* in Lilienthal's birthplace of Anklam, providing another resource for the foundations of this book.

From 2016 to 2019, the *German Aerospace Center* conducted a spectacular research program. At an aerodynamic research institute founded in Göttingen in the early twentieth century, co-author Markus Raffel oversaw a renewed scientific investigation featuring aerodynamic, flight, and mechanical studies of Lilienthal's three basic aircraft designs—his production glider, his large biplane, and his experimental monoplane performed using modern aeronautical research equipment. This research program attracted interest from around the globe. The final part of this book is devoted to the experiments performed as part of this reverse engineering project, offering new insight that would never have been possible without a practical understanding of Lilienthal's approach to flight.

The Sky Over Pomerania

It might be surprising to learn that the search for the roots of human flight leads us not to a big city, not to some prestigious international university, not to a center of technological progress, but to a small town in the north of Germany near the Stettin Lagoon near the Baltic Sea. Why did it take until nearly the dawn of the atomic age for people to glean the secrets of flight from nature? They really weren't even that much of a secret—for thousands of years, birds have been circling in the sky above the heads of envious men, butterflies have been flapping their colorful wings, and bees have been buzzing from blossom to blossom.

> *"One could almost get the impression that the stork was created specifically to incite the desire to fly in us humans, and to serve as our teacher in this art,"*

is how one book described it, today considered the origin of the physics of wings. The book was printed in Berlin, the booming capital of Germany, in 1889, some 41 years after its author was born in the small Pomeranian town of Anklam, near the Baltic Sea. Otto Lilienthal, the first of eight children born to the cloth merchant Gustav Lilienthal and his wife Caroline, came into the world on May 23, 1848 in the shade of the mighty *St. Nicholas Church*, a typical example of the northern German brick Gothic style with a tower looming 103 m above the wide flatlands into the sky, a testimony to the glory of God and the city's pride and wealth. His brother Gustav was just a year and a half younger. Otto Lilienthal remained close with Gustav

M. Raffel and B. Lukasch, *The Flying Man*, Springer Biographies, https://doi.org/10.1007/978-3-030-95033-0_2

throughout his whole life, growing up with him, going to school with him, and watching the storks in the floodplains near the city with him. Gustav helped him make a set of artificial wings from goose feathers, as well as a massive, flapping-wing device with a wingspan of six meters, which they built and tested together. He laid with Gustav in the grass, gazing up at the sky, and wondering why the storks that circled above them were able to glide so effortlessly without ever falling, or even needing to flap their wings.

Three decades would pass between the Lilienthal brothers' forays into nature and their childhood and adolescent inventions before the images of Otto the flying man literally travelled around the world. During this time, the brothers lost their father at a young age, Otto served in the Franco-Prussian War, watched as balloons left behind a besieged Paris, and lived in Berlin as a *bed-renter* who had to share a bed with several other tenants. He studied mechanical engineering at the *Berlin Trade Academy*, later the *Technical University of Berlin*, was hired at his first job, and founded a factory to manufacture steam boilers and small wall-mounted steam engines. After several design iterations, Otto Lilienthal's factory became the world's first aircraft manufacturer in 1893. For the first time in history, flying machines were being regularly produced in series. Built from willow wood and cotton fabric, those early gliders, based on his *Normalsegelapparat* (literally, normal soaring apparatus) design were available for purchase in multiple countries for 500 marks. That's the rough equivalent of six months' salary for an average skilled German laborer of the day, though Lilienthal's customers were likely more well-heeled than that. It's believed that Lilienthal sold nine of the gliders, and surviving examples are displayed in London, Moscow, and Washington, D.C. Another model, one that he used himself, is on display in the *Deutsches Museum* in Munich.

Lilienthal built a dozen or so other designs that he used for testing control methods, performance improvements, and even a flapping wing mechanism. One of these test models, the so-called *Sturmflügelapparat* (*Storm Wing Apparatus*), has survived the test of time—today, it can be found in the *Technical Museum* in Vienna. Our understanding of these other designs of Lilienthal's largely comes from an entirely different invention that Lilienthal encountered while studying the wings of the white stork.

In 1884, the photographer Ottomar Anschütz presented a series of photographs of storks in Berlin, which caused quite a sensation. At the time, small photographs called *carte de visite* were popular items to collect and give as gifts. Given the state of camera technology at the time, photographs required extremely long exposure time, which made taking pictures of things like live wild animals virtually impossible. Anschütz had worked on this

Fig. 1   Ottomar Anschütz: *Storks No. 38*, 1884. © Otto-Lilienthal-Museum. All Rights Reserved

problem for quite some time, and razor-sharp images of the birds nesting and flying proved that he had finally succeeded. He kept the technique secret for years, but, in 1888 patented a high-speed camera whose focal plane shutter was located immediately in front of the light-sensitive photographic plate. This innovation provided the basis for all subsequent mechanical exposure systems and enabled him to capture photos at high shutter speeds, compared to seconds or even longer in typical cameras of the day. This invention enabled him to do what photographers had dreamt of for a half century or so—take pictures of objects in motion.

Lilienthal owned some original prints of the stork photographs. Having chosen the white stork as his "teacher", he was undoubtedly delighted at the ability to study these incredibly clear and precise depictions of the birds' wings in flight. In a nice bit of symmetry, just a few years later, Anschütz photographed Otto Lilienthal during flights near his home on the outskirts of Berlin (Fig. 2).

Lilienthal's flights were no secret in Berlin; he could be spotted by the public almost every weekend after 1894. Lilienthal's *Fliegeberg*, literally *Fly Mountain*, an artificial hill he'd built near his home as a place to test his gliders, became a tourist attraction for Berliners. Today, onlookers would simply pull out their mobile phones and effortlessly record every flight as

Fig. 2 Ottomar Anschütz mounted each of his photographs onto a cardboard frame with the inscription "Nach dem Leben aufgenommen von Ottomar Anschütz" [Captured true to life by Ottomar Anschütz]. Otto Lilienthal in flight on August 16, 1894. ("see Fig. 12, page 122"). © Otto-Lilienthal-Museum. All Rights Reserved

a high-resolution video. And indeed, besides Anschütz himself, several other photographers watched Lilienthal fly and captured his feats on their new and improved cameras. That small group of photographers was generally centered in Berlin, so it's not surprising that the vast majority of photographs of Lilienthal were taken at the *Fliegeberg*—it was only on two occasions that he was able to entice a photographer to visit his "*high performance flying grounds*" in the Rhinower Mountains, a two-hour, 100-km train ride from Berlin. Only a few photos exist of his 250-m flights that area, the last of which, in 1896, took a fatal turn.

The only photographs relating to his accident are two images that show the damaged flying machine in the courtyard of his factory in Berlin, taken days later, likely as part of the police investigation. The circumstances of his final flight are shrouded in controversial eyewitness accounts and conjectures, some plausible and others less so.

The spectacular and striking photographs of the first flying man not only impressed Lilienthal's immediate successors, but most importantly left their mark on the Wright brothers themselves. Generations of pilots have viewed themselves as Lilienthal's descendants, and his imposing personality and breakthroughs in flight inspired more than a few to try it for themselves. Some of the most recent successful examples of those following in Lilienthal's footsteps are discussed at the end of this book.

From Poor Student to Excellent Scholar

"Nothing is better for developing a serious outlook on life than seeing your siblings lying cold and pale, robed in white, surrounded by flowers, in a child-sized coffin."

At a very young age, Otto and Gustav Lilienthal lost four sisters and one brother. Their deaths can be attributed to unsanitary living conditions, sleeping in windowless chambers with dry rot underfoot, and a lack of medical care, according to what Otto later wrote in a family history. Three of their siblings died before age one. The fourth, Wilhelmine, was just four when she died. The two boys were very aware of her death, as they had already developed a close personal bond with her. Otto, the eldest sibling, was born in Anklam on May 23, 1848, with Gustav following suit on October 9, 1849. Regarding his relationship with his brother, Otto Lilienthal later wrote:

"My brother Gustav was and is my second self."

This statement is especially powerful given that Otto wrote it at the age of 46.

M. Raffel and B. Lukasch, *The Flying Man*, Springer Biographies, https://doi.org/10.1007/978-3-030-95033-0_3

"Not only did we share all of our joys and sorrows in early childhood, he wrote, we also undertook all of our foolish pranks and sensible ideas together; not only did we share the same nurturing influence of our excellent mother, but we also advanced toward the same worldview in our ongoing self-education. Many of our biggest undertakings were a joint effort."

In his brother Gustav, Lilienthal saw a

"shining proof that diligence and perseverance in self-improvement provide greater momentum for achieving greatness than early giftedness."

Gustav was only a bad student when forced to struggle with ancient languages in high school. Later, at the newly built middle school in Anklam, he would finally find something that captured his interest, and Otto would undoubtedly have described himself in the same way. He also completed high

Fig. 1 The brothers Otto and Gustav Lilienthal, 1862. *Photo* A. Regis. © Otto-Lilienthal-Museum. All Rights Reserved

school in Anklam with good wishes from his teachers than praise for his academic accomplishments.

An image of Otto and Gustav as young adolescents shows just how markedly they differed in appearance. In the only known picture from this period, most likely taken after the death of their father, Otto is around 14 years old, and Gustav is 13. Both are dressed in the same way. They are wearing light-colored trousers, checkered tunics with a small white collar, and broad leather belts around their waists. Gustav is sitting in a wicker chair, and Otto is standing tall beside him, beaming into the camera. Gustav, on the other hand, is looking somewhat shyly and suspiciously at the viewer, with his hands lying in his lap and his feet awkwardly placed. Otto—the older brother—projects a more dominant appearance, wrapping his arm protectively around his brother's shoulder, as if to take his father's place. Many of the details of his life story may have been lost over the years, and Gustav never fully managed to emerge from his brother's shadow, even in old age. When it came to ingenuity and the inclination for invention, the brothers seemed perfectly matched. But Otto stood out for his athletic performance and bravado, whereas Gustav himself admitted that those were never his strong suits. As adults, Otto remained significantly taller, with curly blond hair, the perfect image of a constantly happy, optimistic, and successful man. He was always friendly and high-spirited even when facing financial and other pressures. Descriptions invariably paint him as a sociable, capable, and hands-on individual with energy and drive, who enjoyed entertaining the ladies. Gustav, on the other hand, had a quieter presence, one that was more prone to pessimism and melancholy.

Otto Lilienthal himself told the story of his ancestors, parents, and siblings in a family history written in 1894 or 1895, not long before his death. His son Fritz then carried on the family research for decades, continuing his father's records.

Most of Lilienthal's ancestors were farmers living in Western Pomerania, west of the Oder River, which was a Swedish territory until the beginning of the nineteenth century. Lilienthal described his forebears as a healthy and extremely proud lineage. According to Lilienthal, almost every man among his forefathers was possessed of excellent physical strength and great agility. He also wrote that a special talent for mathematics ran in the family. His father even wrote a textbook on arithmetic, a project that was far removed from his profession as a cloth merchant.

Everything that Lilienthal wrote about his ancestors could be applied to the man himself. His son Fritz reported that all of his teeth remained present and healthy until his death. He was six feet tall with an impressive figure and extraordinary agility.

Lilienthal inherited his father's passion for mathematics and technology, which formed the basis of his success as an outstanding mechanical engineer. From his mother's side, he inherited remarkable artistic talent. Together, these qualities allowed him to take on his prominent role as a creative inventor, especially in the development of aviation.

Otto describes his memories of his father in the family chronicle:

"I was a boy of 13 when I lost my father, so my personal memories of him tend to relate to physical qualities. I know him as a skilled swimmer and diver who considered it beneath his dignity to enter the water from a high diving board without performing a somersault. An obsession with bravura must truly be a Lilienthalian trait. My father would repeatedly throw his wedding ring into the Peene river and dive to the bottom to retrieve it, until one day it was finally lost in the mud. When I was barely 5 years old, he took me on his shoulders and jumped into the water with me. It is little wonder that I would later become a passable swimmer and diver myself. My father's scientific interests focused in particular on mathematics. Politically, my father always expressed democratic tendencies, and supported whatever progress would bring the most freedom."

Otto's mother also described her husband lovingly:

"When I met him personally, I realized that he was not indifferent to me. We were soon seeing each other every day. I found him a pleasant man, though very quiet in company. Perhaps my own presence is the cause of his silence. I must admit that I had not noticed him before, but I found his reserved disposition toward me very pleasant."

But the character of Otto's father was undoubtedly more problematic than suggested by these generous recollections. He left his future wife and her family in the dark about his financial situation. When buying a house on the Peenstrasse near the town's market square, it appears that Otto's father overextended himself financially, running up considerable debt, which he then sought to conceal and deny. His future mother-in-law, a respected citizen in the small town, found out and attempted to prevent the marriage. As a result, Otto's parents were married against the wishes of his grandmother. A few years later, the house had to be resold, as Otto's father's dire financial straits forced him to file for bankruptcy. With help from relatives, Otto's mother managed to buy a smaller house in her own name at the lower end of the same street, near the river. At that point, another of Otto's father's character traits came to light: faced with increasing difficulties, he began to retreat more and more frequently to the local inn to spend his time playing cards and debating politics. At the time, support for the revolutionary events unfolding in Berlin was growing, with plans to introduce universal voting rights and abolish the privileges of the nobility.

When Otto's father Gustav Lilienthal heard about the growing violence in the capital, he immediately wanted to travel to Berlin to participate. He was only dissuaded from doing so, with great difficulty, because Otto's birth was imminent. The German revolution of 1848 climaxed in a constitutional assembly in St. Paul's Church, but ultimately lost its momentum. The nobility regained their powers and punished those who had advocated for reform. For the merchant Lilienthal, this meant losing his wealthy customers; soon afterward, he was forced to relinquish a peat-cutting site that still provided the family with some income.

Several years later, like many others who had opposed the nobility during the revolution, the family decided to improve their lot by emigrating to America. Otto's mother Caroline was prepared to take this risk together with her husband and was enthusiastically on board. But Otto's father died while the plans were still being drawn up. He left behind his 12-year-old son Otto, his 11-year-old son Gustav, and his daughters, Marie and Anna, who were only four years and three months old, respectively. Otto's loving description of his father in the family chronicle shows that his mother and grandmother only ever spoke well about his father, out of love for the children and the man himself. But they also relentlessly enforced a strict prohibition against playing cards or drinking alcohol. Even as adults, neither Otto nor Gustav reportedly ever touched playing cards.

Otto's father Gustav attended high school in Stralsund, where he learned the cloth merchant's trade. He put down roots in Anklam by purchasing a grandiose place of residence and business in a prime location near the large market square in the city center, next to *St. Nicholas' Church*, which dominates the cityscape. Otto's parents were married there in 1847.

Otto's mother's maiden name was Caroline Pohle. Her father came from Berlin and worked as a military doctor in Stettin. Thanks to wealthy relatives in Dresden, Caroline was able to study singing with teachers in Berlin and Dresden. Even after she married, Caroline continued to sing frequently in concerts in Anklam, and was very successful as a singing teacher. She awoke an interest in both art and science in her children.

"Our house was a hub for artistic activity within our hometown,"

wrote Otto Lilienthal.

Fig. 2 The Lilienthal family's residence and place of business. *St. Nicholas' Church* on the market square, where the children were baptized, can be seen in the background (photograph taken after 1910). © Otto-Lilienthal-Museum. All Rights Reserved

When Otto's father died of an acute lung disease in 1861 at the age of 44, his brother Wilhelm Lilienthal, an estate tenant, became the children's guardian. Wilhelm was married to Mathilde Pohle, the sister of Otto's mother.

This marked the start of a difficult time for the family. Caroline Lilienthal was deeply devoted to her children, and worked tirelessly to support them, ensuring that all three would get a good education. She died in 1872 as Otto and Gustav were preparing to take her to Berlin with their sister Marie.

"The love and respect of the citizens of Anklam was on full display at my mother's funeral in February 1872, as the long funeral procession travelled over flowers

scattered on the road, stretching from our house far into the churchyard. Our mother died shortly before we were able to move her to Berlin to repay her for what she had done for us. She had already sold our little house in Anklam, but only our grandmother and sister could join us in Berlin."

Otto's only remaining siblings were his brother Gustav, and his sister Marie, who was eight years younger. He continued to care for them attentively until the end of his life. After their mother's death, Otto supported his sister as she trained to become a teacher. When Marie later married a farmer in New Zealand and ran into difficulties, he toyed with the idea of bringing her and her family back to Berlin.

As for his brother Gustav, Otto shared more than just childhood interests with his *"second self"*, as he'd described him. Even today, it is difficult to say which brother was responsible for each of their ideas and patents, as they developed many of their inventions together. As a precautionary measure, Otto, while working for the Hoppe company in Berlin, registered his patents under his brother's name. And many of Gustav Lilienthal's spectacular toy patents list Otto as the original inventor. Their lifelong interest in flying was always something that brought them together. The subtitle of the book *Birdflight as the Basis of Aviation* even states:

Compiled from the Results of Numerous Experiments made By O. and G. Lilienthal.

Long after Otto Lilienthal's death, Gustav continued his brother's work with aircraft, eventually passing away himself at the age of 83 from a heart attack next to his huge *Grosser Vogel* (Big Bird), a motorized flapping-wing machine built at the airfield in Johannisthal that never successfully left the ground (Fig. 3).

Otto wrote about the brothers' talents in the family chronicle:

"In our childhood, people thought we were opposites. I showed an early talent for drawing, modelling, and carving, and people always thought I would be an artist. Lacking any other special talents, my brother's only prospect was a commercial profession. How remarkable therefore that, of the two of us, my brother would become the artist, while I would throw myself into the arms of technology."

Both brothers attended high school in Anklam on the Wollweberstrasse, Otto from Easter 1857 to autumn 1864, Gustav from autumn 1858 onward, and both students struggled with Latin. It took Otto seven years to complete secondary school from grade five to grade nine, as he was held back twice.

Fig. 3 Gustav Lilienthal's Grosser Vogel, Berlin-Tempelhof 1928. © Otto-Lilienthal-Museum.

When he transferred to the provincial trade school in Potsdam, his professor had little to say about him other than to wish him good luck. According to his teachers, the only area in which he showed interest or skill was drawing, and achieved only satisfactory results in mathematics and gymnastics.

This makes it all the more astonishing that, merely two years later, Lilienthal passed his exams with the best ever results recorded at the trade school in Potsdam: "excellent" in every subject. His diploma was awarded with distinction. In addition to his talent, the faculty explicitly praised his work ethic.

According to his brother, Gustav similarly struggled with ancient languages in high school. Gustav finally found something that captured his interest when he attended the newly-built middle school in Anklam. He became a bricklayer's apprentice and later a guild-certified journeyman. After successfully passing his exams, he followed his brother's example by registering as a student at the Building Academy in Berlin, and worked as a construction manager in Berlin, Prague, and London. When he returned to Berlin in 1874, he once again took part in Otto's experiments researching and documenting the properties of artificial wings and the fundamentals of flight.

The Lilienthal brothers' passion for flight began in their childhood. According to Gustav, they were captivated by a popular youth publication at

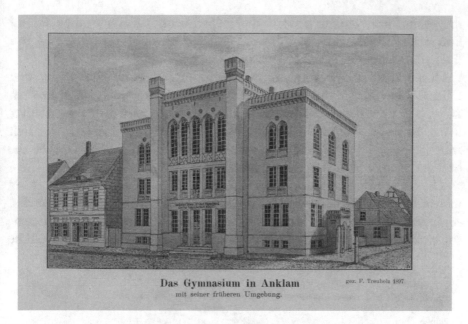

Das Gymnasium in Anklam
mit seiner früheren Umgebung.

gez. P. Treuholz 1897

Fig. 4 The new high school building in Anklam. It was inaugurated in 1852 by King Friedrich Wilhelm as a model for a modern school building in Prussia. © Otto-Lilienthal-Museum. All Rights Reserved

the time about the travels of Count Zambeccari, an Italian aviation pioneer who ultimately lost his life while ballooning. On the peaceful forest meadows of the Karlsburger Heide, across the river and not far from their house in Anklam, the brothers observed storks in flight:

> "*We often crept up very close, even with the wind, as the birds had a very poor sense of smell. If a bird suddenly caught sight of us, it would leap up, hopping toward us, until the force of its wings was enough to lift it away from the ground. At this point, we already understood that taking off against the wind must logically be easier than with the wind, because such a shy animal would not jump toward danger for no compelling reason.*"

Decades later, Otto acquired a few young storks of his own, kept them in his yard and watched them as they learned to fly.

When the brothers were 13 and 14 years old, they built their first pair of wings—Gustav alleged that they had their mother's permission. The wing surfaces were two meters long and one meter wide, made from thin beechwood boards. The underside was fitted with straps for the boys to put their arms through. Their plan was to run down a little hill, flapping their wings against the wind like a stork. To avoid being ridiculed by their schoolmates, they waited until dark to perform their experiments. Of course, they did not find much success, with a lack of wind beeing the least of their problems. But their youthful enthusiasm survived unscathed.

During their time in Anklam, they performed the early experiments that Lilienthal included in his book *Birdflight as the Basis of Aviation* more than 20 years later. In the attic of their house, and later in a barn owned by relatives living nearby, the brothers installed large man-carrying devices to research wing flapping. Their last and largest installation was tested in 1868 when the brothers were at home on holiday from school.

The contraption consisted of two pairs of wings with a wingspan of six meters that could be moved in opposite directions with the leg muscles, equipped with flaps that opened on the upstroke and closed on the down-stroke. The brothers used counterweights to measure how much lift could be

Fig. 5 Description of the experimental apparatus built to research wing flapping in Lilienthal's book, published in 1889

generated by flapping the wings as strongly as possible. Lilienthal described the results:

> "*After a little practice we were able to lift half our weight, so that a 40-kilogram counterweight sufficed to just balance the machine and operator, weighing together 80 kilograms. The requisite effort, however, was so great that we could maintain ourselves at a certain level only for a few seconds.*"

The amount of lift they were able to generate, 40 kg, was not enough to take off from a standing position by flapping the wings. Vertical takeoff was clearly impossible through muscle power alone.

There had to be another solution.

Lodger–Engineer–Manufacturer

After attending the provincial trade school in Potsdam thanks to help from his mother's relatives, Otto moved to the Prussian capital of Berlin in the autumn of 1866. In Berlin, the dawn of the age of technology could be both seen and felt. The city was transforming into an industrial metropolis at breakneck speed. The walls of the city were razed, and less than two kilometers away from the royal palace, a company called Borsig built more locomotives than in the rest of Germany combined. The Berliners had a word for the factory district of their growing industrial city—*Feuerland*, the *land of fire*. In the factory district, the chimneys of the iron foundries and various factories colored the façades of the houses grey, and the steam hammers rattled their windows. To accommodate the vast sea of workers pouring into Berlin, early tenements were built right next to the factories. Each house had multiple backyards. The original idea was to prevent the formation of slums as found in other large cities by uniting every class of the population into a single house with a common entrance. The large gates at the entrance led to numerous yards and wings. The street fronts featured elaborately designed and richly decorated façades, with tall and distinguished upper-class apartments behind them. But the farther back one ventured, the more precarious the living conditions became. The far end of the tenement building, away from the

Supplementary Information The online version contains supplementary material available at (https://doi.org/10.1007/978-3-030-95033-0_4). The videos can be accessed individually by clicking the DOI link in the accompanying figure caption or by scanning this link with the SN More Media App.

M. Raffel and B. Lukasch, *The Flying Man*, Springer Biographies, https://doi.org/10.1007/978-3-030-95033-0_4

street, held bare apartments with nothing but a single room serving as both kitchen and living space, with no sanitary facilities. The same apartment would often by shared by multiple generations.

But Otto was enthralled: a new era was beginning—an age of technology that would allow every problem of the past to be solved, even social ones. His goal was to study mechanical engineering at the Berlin Trade Academy, the forerunner of the technical university that still exists in Berlin today. Having received high marks and his diploma from trade school, he had already fulfilled most of the eligibility requirements, but he still needed an internship. And so, he began working as an intern at an iron foundry and factory owned by legendary industrialist Louis Schwartzkopff on the Chausseestrasse. At the time, the company manufactured railroad switches, turntables, and water systems for railways. A year later, it also began building locomotives.

Lilienthal later told his friend, a Berlin meteorologist named Carl Kassner, about his living conditions during his first year in Berlin:

> "*I rented a place to sleep together with a cab driver and a drayman. The cab driver worked nights, so I only had to share the bed with the drayman. I bought rye bread every day from a disabled man sitting in front of the Oranienburger Tor.*"

Lilienthal's fate was shared by many other young people who had to share their bed with several other tenants. For many Berliners living with their children and grandparents, subletting a bed like this was the only way to survive.

During his internship, Lilienthal showed great skill in all kinds of precision work. He was given the opportunity to work in the engineering office, not only giving him the position, he'd secretly hoped for, but also bringing a welcome increase in wages over those of a lowly intern. In the autumn of 1867, he enrolled in the *Trade Academy*, opened in 1820 by King Friedrich Wilhelm III, who dedicated it to industry.

Lilienthal studied mechanical engineering for six semesters. His most prominent teachers included the director, Prof. Franz Reuleaux, who taught mechanical and other types of engineering—crucially important topics for Lilienthal—and the mathematician Erwin Bruno Christoffel, who is well known to this day for introducing concepts that eventually helped provide the basis for general relativity. Reuleaux himself is remembered as a pioneer of scientifically driven mechanical engineering. He was a member of a commission of experts established by the Prussian government in 1867 to be entrusted with the task of designing a program "*for experiments intended to determine the laws of air resistance for the manufacture of dirigible aircraft such as those currently being used*".

Since Lilienthal's burning interest in flight was well-known to his fellow students and his teachers alike, it seems likely that Prof. Reuleaux was also aware of it. Reuleaux, who made ground-breaking contributions to the technical sciences, appears to have followed Lilienthal's subsequent career with interest and concern.

Prof. Christoffel, the mathematician, also knew of Lilienthal's interest in flying and never objected to it, according to his student's later recollections. Studying the topic could surely do no harm, but Christoffel warned against investing money into such matters. At the time, the general and scientific consensus tended to see greater prospects for flight in airships rather than heavier-than-air aircraft. Thanks to support from Christoffel and Reuleaux, Lilienthal was later awarded a scholarship of 200 Thalers (a one-ounce silver coin) per year for two years. Otto took advantage of this windfall as an opportunity to bring his brother to Berlin, allowing the latter to enroll at the Building Academy, which later also became part of the technical university.

Otto passed his final examinations in late July 1870 with very good results, two weeks after the outbreak of the Franco-Prussian War. In most subjects, he received the grades *quite good* and *very good*, with four grades of *excellent*, including in Prof. Reuleaux's subjects, much to his delight. He, along with many of his fellow students, then immediately joined the Prussian Guards Fusilier Regiment as a one-year volunteer to complete his mandatory service. Lilienthal participated in the siege of Paris and retired at the end of his year as a non-commissioned officer. Gustav did not complete his studies with a diploma because the school was closed when the war broke out. That said, Gustav was lucky, in a sense, in that he avoided conscription for health reasons.

———

For his first job as a young engineer, Otto started work in the small factory of M. Webers in the mechanical engineering district of Berlin in 1871. The factory was housed in a former dance hall and was managed by Emil Rathenau. Rathenau took over the workshop in 1865 and managed it successfully for almost a decade near the famous factories of Egells, Borsig, Wöhlert, and Schwartzkopff. The factory's primary business was manufacturing steam engines. Rathenau later founded the *Allgemeine Elektrizitäts Gesellschaft* (AEG), which became part of the *Daimler-Benz AG* in 1995.

After just one year, in 1872, Otto changed jobs to become a design engineer in the engineering workshop owned by Carl Hoppe on the adjacent street, the Gartenstrasse. Over the next few years, at the behest of Hoppe, who manufactured mining machinery, Lilienthal travelled to every major mining region in Saxony, Silesia, and Galicia, a part of Austria-Hungary at the time. He also designed his own lightweight cutting machine for mining coal and salt, applying for his first patent in 1877, strategically choosing to file it in

Fig. 1 Lightweight hand-operated cutting machine for coal mining, designed by Otto Lilienthal while working as an engineer. Patent No. 4771 was granted for this machine in 1877 in the Kingdom of Saxony. © Otto-Lilienthal-Museum. All Rights Reserved

the Kingdom of Saxony. It was technically his brother Gustav who applied for the patent, presumably to clearly separate it from Hoppe. Gustav also worked underground with Otto to test the machine in the *Royal Saxon Colliery* of Potschappel near Dresden. Later, he sold several machines for use in the salt mines of Wieliczka. The mines are now part of Poland and are designated as a *UNESCO* World Heritage Site. Besides this economic success, Otto's time in Saxony had a significant impact on his personal life—on June 11, 1878, he married the daughter of his mining foreman.

At the time, Otto and Gustav were living with their sister Marie and their grandmother Pohle on the Albrechtstrasse near the Spree River in Berlin. Gustav was also frequently on the road, doing temporary work as a builder in Prague and London. In 1874, Gustav returned to Berlin and performed a series of aeronautical experiments to quantify the properties of curved wing surfaces.

Even though he did not earn a university degree, Gustav became a creative architect, civil engineer, artist, and educator. Even today, we cannot clearly separate each of the brothers' contributions to their inventions. The only patent that lists them jointly as inventors is *United States Patent Office* No. 233.780. *Composition Toy Building-Block*, a US-based spin-off of possibly their most momentous invention besides their gliders. Their toy blocks

Fig. 2 The Lilienthal siblings, 1872, Otto (24 years old). © Otto-Lilienthal-Museum.

Fig. 3 Gustav (23). © Otto-Lilienthal-Museum.

Fig. 4 Marie (16). © Otto-Lilienthal-Museum. All Rights Reserved

would later conquer the world as *Anchor Stone Blocks*, the first of many construction-based toys that remain popular to this day.

After Otto's wedding, the couple, along with Gustav, moved into a three-room apartment on the Brunnenstrasse in Berlin. It wasn't long until their apartment was transformed into an inventor's workshop. The kitchen was used to experiment with baking artificial sandstone, pressed together from sand and varnish, no doubt wreaking havoc on the young family's sense of domestic tranquility. In 1879, Otto's son, the family's first child, was born in this apartment.

The lack of economic success with the stone building kit prompted the brothers to sell the idea to Friedrich Adolf Richter, which led to a lengthy legal dispute five years later. This was no doubt one reason why Gustav was looking for a new start. He was also uncomfortable with living as a subtenant while his brother's family continued to grow. And so, it was Gustav who would ultimately pick up where his parents left off by emigrating to find a new beginning in the "*new world*". His parents had chosen the USA as their destination, but Gustav's path led him to Australia.

Fig. 5 Drawings by Gustav Lilienthal, as found in his building kit, which would later be developed into the globally popular *Anchor Stone Blocks*. © Otto-Lilienthal-Museum. All Rights Reserved

Fig. 6 Gustav Lilienthal's model building kit, the forerunner of many modern construction toys. © Otto-Lilienthal-Museum.

> *"In 1879, we invented a building block kit from a varnish-chalk compound. In 1880, we sold the recipe to Richter in Rudolstadt, who manufactured countless millions of building blocks with it,"*

Otto wrote. To avoid further legal disputes, Gustav later switched to a different material and system. His model building kit with perforated wooden strips was a precursor to many other building kits that were later made from metal and plastic. It was also patented under Otto's name, even though Gustav was the primary inventor.

In 1878, Marie moved to Ireland to be a teacher. In 1880, she joined Gustav on a trip to Australia, and Gustav found professional success in

Melbourne. Marie married a farmer and moved to New Zealand. After five years, Gustav returned to Germany, intending for it to be a vacation. But he ultimately decided to turn his back on Australia forever, a decision that was likely heavily encouraged by Otto.

—

Otto filed a patent for steam boiler innovations in 1881, but it wasn't until 1883 that he established his own workshop to manufacture the new, safer

Fig. 7 The Lilienthal *Small Motor* consisting of a safe, explosion-proof coiled tube boiler and a lightweight wall-mounted steam engine: the founding basis of Otto Lilienthal's company. (▶ https://doi.org/10.1007/000-6yg) © Otto-Lilienthal-Museum.

boilers and lightweight steam engines. His workshop was on the Köpenicker Strasse, in a new and growing industrial district in the eastern part of the city. The workshop opened with just two vices and a lathe, but as customers began to place orders, it soon turned into a small factory. Otto had at last achieved his long sought-after self-sufficiency.

The family was now just down the road from the factory. This is where Otto's daughter Anna was born in 1884, followed by his son Fritz in 1885. The couple's fourth child, Helene, was born in 1887 in a house they owned in the Berlin suburb of Gross Lichterfelde. The house, which featured a yard and workshop for flying machines, was designed by Gustav, like many of the other houses in Lichterfelde. Today, many of these houses are recognized as historical monuments.

When Otto died in 1896 at the age of 48 following his crash, his children were 17, 12, 11, and nine years old. His wife Agnes outlived him by 24 years, his sister Marie by 16 years, and his brother Gustav by 37 years.

Life Goal: Inventor

"I hope that at some point Gustav and myself will be in a position to settle down somewhere without any specific business and focus on all kinds of inventions, especially flying. We naturally cannot expect to earn much from this at first; one must live on one's interest. That is roughly the ideal that we will be striving toward."

Otto wrote the above words to his sister Marie in New Zealand in 1885, two years after opening his own factory. He makes it seem as if his very successful factory, which had been granted patents and achieved national recognition, was simply a means to an end to allow him to work on *"all kinds of inventions, especially flying"*. But this was most certainly not true. Lilienthal was quite capable of pursuing multiple goals simultaneously with his full attention.

He undoubtedly corresponded intensively with his brother in Australia, but their letters have not survived. Just as Otto wrote to his sister, he would have strongly encouraged and supported Gustav's decision to return to Germany. Given the lengthy legal dispute over the brothers' original building block set, and the difficult start with the new one, Gustav may have had doubts about returning to Germany.

But Otto missed the creative duality that he shared with his brother. He wanted to continue building on their shared inventions from bygone times. Though entirely dissimilar, the brothers needed each other, and complemented each other very effectively. Otto had learned some difficult lessons

M. Raffel and B. Lukasch, *The Flying Man*, Springer Biographies, https://doi.org/10.1007/978-3-030-95033-0_5

about asserting himself in business. Being a strict employer and a shrewd busi-nessman were not his life goals. But they were a prerequisite for the freedom to work on bigger things like flying and trying to solve the great questions of society—and this at a time where industrialization was just beginning to unleash social upheaval in Germany. He shared this approach to life with his brother.

Gustav had become engaged, but, for two years, he only dared to write to his fiancée in secret, arranging for his letters to be collected at the post office. He had not yet managed to establish himself in Germany. How, then, could he presume to ask her father, a respected doctor, for his daughter's hand in marriage?

In one of his secret letters, he wrote:

"If you should ever happen to mention the state of my circumstances and my plans, please emphasize that it is not my intention to restrict my activities to manufac-turing toys forever, I simply wish to secure funds to pursue other ideas. I still consider that my primary task in life is to build small, healthy houses for families. For me, this project is not simply a source of income but will provide a timely redress of many grievances that account in no small part for the general dissatisfaction of all working classes."

Both brothers remained true to their goals. Their many inventions did not bring them great wealth. But both Otto and Gustav came very close to accomplishing their goals in life. Gustav did build houses in green spaces that continue to be admired today as listed heritage-protected buildings. On behalf of the Bodelschwingh Foundation, which still exists to this day, he built the first ever homeless shelters to offer dignity to their residents, an accomplishment for which he is still remembered and honored. Houses based on Gustav's designs were built for the destitute in a vegetarian colony named after the biblical garden of Eden, and there, too, he is remembered as a pioneer. He also built a housing co-op called *Freie Scholle* (free soil) in the northern part of Berlin, which also still exists. A memorial to Gustav was erected on the site—remarkably, this happened in 1945.

There were plenty of ways for Otto to spend the money he earned. Flying was just one of them—the theater was another. Otto became a co-owner of the *Ostend Theater* in the less fashionable eastern part of the city, grimly described by Berliners as the "mass grave of the Far East" due to its dire finan-cial straits. He transformed it into the *National Theater*, an inexpensive stage for the people, while trying his hand at being an actor and author himself. The theater's goal was to bridge the chasm between the rich cultural life in central Berlin and the everyday lives of its factory workers. He also distributed

Fig. 1 Otto Lilienthal as an actor, 1893. © Otto-Lilienthal-Museum. All Rights Reserved

theater tickets to his workers. It was only a 30-min walk to the theater from Lilienthal's factory.

Otto wrote a play, *Moderne Raubritter (Modern Robber Barons)*, pouring a lot of his own life into the text.

"Gentlemen, I'm sure you believe that I would do anything within my power to prevent such a large business opportunity from slipping through my fingers. There is no way to make savings on wood and metal fittings, so I would have to reduce

my people's pay, but I would rather refuse your order than become an oppressor to my workers. [...] A trained, reliable worker should earn 25 to 30 marks a week. That's the bare minimum that a working-class family needs. If I cannot ensure that my people earn this much, then I must be thankful for being an employer."

That line is spoken in the play by a young factory owner to his business partner in the first act. Krüger is a furniture-maker whose character is suspiciously similar to Otto's own and who lives his life according to the same beliefs. Another episode of the play contains other parallels to events in Lilienthal's personal life. Krüger meets his future bride while singing, just as Otto and his own wife sang together in a pub in the mining village of Potschappel.

Lilienthal was also making history as an employer in the real world: he is thought to be the first entrepreneur in Berlin to introduce profit sharing within his company.

"To improve the interest of my workers in my business and to offer them the opportunity to increase their income by contributing according to their own performances, I intend to discontinue piecework and introduce participation in the net business profits, initially in the amount of 25%, while retaining the current wages and factory ordinances."

He made this announcement on March 12, 1890, in his factory. The measure is said to have remained successful until his death. Given the abundance of mechanical engineering businesses in Berlin, Lilienthal was constantly fighting to keep his core workers from leaving for one of his many competitors.

Lilienthal later attributed his success as a machine manufacturer to his aeronautical work, no doubt with some satisfaction. The idea to equip an aircraft with a motor led him to design a very small engine powered by a miniaturized pressure boiler. This design had been sitting on his desk ever since model testing. Now, Otto recognized a need in the market that could be met by a larger version of the pressure boiler. Otto's design was lightweight and safe, making it ideal for the many small factories and workshops of Berlin. These shops were often located in the *Souterrain*, an elegant French word borrowed by Berliners to describe the expanded basement floors in the tenements, or on the ground floors of residential buildings. Due to a large number of boiler explosions that claimed multiple victims each year, the installation of traditional steam boilers in residential buildings had been prohibited by the *Prussian Steam Boiler Act*.

Lilienthal initially commissioned a small workshop on the Linienstrasse in Berlin to produce the parts that could not be built in his own attic workshop. His calculations were sound. The difference between Lilienthal's coiled tube design and other boilers was that the steam was not generated and saved within a conventional boiler, but inside the coiled tubes, which were filled with a mixture of water and steam. The narrow pipes made dangerous explosions virtually impossible, even under the most unfavorable conditions. Because of this, there was no objection to using them in residential buildings, which opened a substantial market for Lilienthal.

Otto had opened his factory in 1883. Business quickly picked up and, after 10 years, he had about 60 employees and was greatly valued as an employer.

Fig. 2 Advertisement for the safe steam boiler made by Lilienthal's factory. © Otto-Lilienthal-Museum.

An apprentice later reported his impressions about the factory as follows:

"*Otto Lilienthal was admired by all of his workers. He exuded a kindness and philanthropy that opened the hearts of his employees to him. He had a kind word for everyone, and he and his workforce seemed like one big family.*"

But the workers' thoughts about Lilienthal's flight tests are also interesting:

"*As much as Lilienthal was recognized by his workers as a capable mechanical engineer and expert, the same workers dismissed his attempts to fly in a disparaging, almost mocking manner. One old drill operator even described it as arrogance and defiance in the face of God that could only be met with punishment.*"

—

For Gustav, the elegant English Tudor-style suburban villas known as the *Castles of Lichterfelde* that he had designed in the 1890s became a signature trademark. He oversaw a large number of construction projects simultaneously. His own house, however, was unusually small, but, even so, he rented out two of its rooms. For his brother, he designed a house with a completely different style, also in Lichterfelde—a pavilion with a yard and a large workshop. A platform was later added to the yard to serve as a springboard for early flying experiments (Fig. 3).

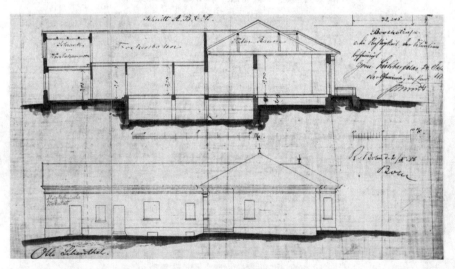

Fig. 3 Otto Lilienthal's country house, built in Gross Lichterfelde, a suburb of Berlin, in 1886. The house was designed by Gustav Lilienthal, and the construction was entrusted to the master mason W. Ernst from Steglitz, who had attended the Building Academy in Berlin together with Gustav. All five rooms were equipped with in-floor heating. The yard was extended with a large workshop. © Deutsches Museum. All Rights Reserved

The Berlin suburb of Lichterfelde was a model project, a kind of reformist estate. The merchant, entrepreneur, and urban developer Johann Anton Wilhelm Carstenn had purchased a manor on the outskirts of the expanding city of Berlin with the goal of building a colony of villas after the English model—a garden city with communal buildings, neatly arranged streets and

Fig. 4 Villas designed by Gustav Lilienthal in the English Tudor style, 1893. © Otto-Lilienthal-Museum. All Rights Reserved

squares, and last but not least a tramway. Carstenn's success with this project even came to be reflected in his name: Emperor Wilhelm I granted him a title of nobility—*von Carstenn-Lichterfelde*. Many of the Lichterfelde villas built by Lilienthal can still be visited and admired today, now protected as historical landmarks (Fig. 4).

—

In addition to his safe steam boilers, Otto's factory manufactured wrought iron pulleys and complete transmission systems, as well as heaters and polyphonic foghorns that could be used as warning sirens on coastal shipping routes.

It is worth noting that the coiled tube boiler was featured in 1882 by the journal of the *Verein zur Förderung der Luftschifffahrt* (VFL), generally translated now as the *Society for the Promotion of Aeronautics*, as a conceivable way to build small engines as a propulsion system for an airship. Over the next few years, Lilienthal improved the details of the design, registering patents for his innovations domestically and abroad (Fig. 5).

By Easter 1884, the inventory of Otto Lilienthal's factory listed assets worth 16,000 gold marks. The next year, he wrote to Marie that his factory was still growing. He already had the capacity for 36 workers, totaling sales of 54,000 marks over the first five months of 1885, almost twice as high as the previous year.

As a factory owner, Lilienthal became a member of the *Association for the Promotion of Industry in Prussia*, in which his former professor Franz Reuleaux and fellow student Adolf Slaby, later the director of the Trade Academy, both played leading roles. In 1890, he gave a lecture on the feasibility of free flight to the association. But beyond this lecture, he did not make much of an impression within the association, as opposed to his extensive participation in the VFL in Berlin.

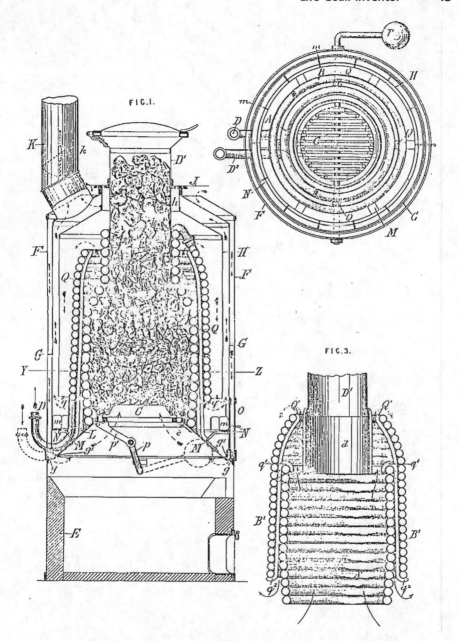

Fig. 5 Illustration of the coiled tube boiler in the English patent granted to Lilienthal. Otto, Bashall, William: *An Improved Coil Steam Generator*, Brit. Patent No. 8322 of June 23, 1886. © Otto-Lilienthal-Museum. All Rights Reserved

From the Dream of Flight to the Science of Aerodynamics

After returning from the Franco-Prussian War, Lilienthal first began to experiment with small flapping-wing models. In 1873, he gave his first public lecture on flight to the *Potsdam Trade Association*, speaking about the theory of bird flight. He was likely invited to deliver this presentation by Slaby.

The young engineer from Berlin presented some surprising claims to his audience of civil servants, entrepreneurs, and engineers. He asserted that the current accepted approach to aviation, which was essentially synonymous with balloons and airship travel, was wrong. A lot of money had been poured into airship technology, which had shown promise in the Franco-Prussian War and was expected to be useful for military, scientific, and civil purposes like carrying the mail. If air transport was to become as viable as shipping and rail, the general consensus was that it would be through the use of dirigibles. But Lilienthal disagreed.

His audience would probably not have known that his ideas directly contradicted the world-famous Berlin physicist Hermann von Helmholtz. Just a few months earlier, on July 26, 1873, at a general meeting of the Royal Prussian Academy of Sciences, von Helmholtz had claimed to have proved mathematically that, "*even with the most skillful, wing-like mechanism operated using their own muscle power*", humans could never achieve "heavier-than-air" flight. According to von Helmholtz, nature had already reached its limits in the great vulture.

Lilienthal's lecture argued that, contrary to the prevailing consensus, dirigibles and other airships were not the answer to the problem of flight—rather,

M. Raffel and B. Lukasch, *The Flying Man*, Springer Biographies, https://doi.org/10.1007/978-3-030-95033-0_6

they were merely an obstacle on the path to human flight. To demonstrate a *heavier-than-air* aircraft, he presented one of his small flapping wing models in flight to his audience.

Just as the Wright brothers realized two decades later, Lilienthal understood that the question of flight would require theoretical tools that did not yet exist. In 1889, he wrote the following passage in his book:

> "*Knowledge of the mechanical processes involved in bird flight currently remains at a level which does not befit the general state of science. [...] At the very least, our knowledge of the laws of air resistance is still so poor that any attempt to approach the problem of flying with calculations necessarily lacks the proper foundations.*"

In 1874, Otto and Gustav began to systematically research the effect of forces on wings of various shapes. Studying the wing shapes and properties of the white stork, and incorporating key properties of its wing surfaces, such as the curvature and shape, into artificial wing models produced measurement results that are still used to describe wings to this day. The Lilienthal brothers described wings in terms of two forces, which they named *lifting* air resistance and *inhibiting* air resistance. Today, we use the terms lift and drag. The goal of a wing is to maximize the lift and minimize the drag. Lilienthal summarized this with an astute formulation:

> "*All flying relies on creating air resistance, all flying work involves overcoming air resistance.*"

By 1874, using rotational devices to study the properties of various curved instead of flat wing profiles, or airfoils, Lilienthal had developed significant insight into the young science of aeronautics. The results of these experiments then gathered dust for more than a decade, and weren't published until 1889. For the 15 years in between, the brothers were kept busy by their professional careers, and lived apart for five of those years. Otto only resumed his aeronautical work in 1888, but not before he had checked his earlier results once again with improved and larger-scale measuring equipment. Lilienthal justified his approach in a letter:

> "*The contents of my work contain so much that is new and different from the conventional assumptions and viewpoints that I could expect the contradictions to be many and various from the very start. But this was also why I chose not to disclose my findings to the public until the entirety of the materials could be presented in a more finished form, and, in my view, the consistency of each result followed from the others.*

The key results of our research had already been established for five years, after devoting the majority of our free time since our academic studies to the problem of flight for some twenty years, when I made arrangements in my newly built home at Lichterfelde near Berlin to reproduce the same fundamental experiments with new, improved equipment, on a large scale, and in a more useful context in collaboration with my brother, so that we would have the sharpest possible control over our previous findings."

Lilienthal's restraint was not unjustified. He was committed to decisively and definitively distinguishing himself from others who were also studying the problems of manned flight. In one article, he wrote:

"The unedifying polemics written by aeronautical hotheads have drowned the specialist literature in an ocean of unfruitful notions. Ignorance, righteousness, and conceit have soured the entire body of aeronautical literature with a bitter after-taste [...] To make things worse, laymen add their emotional theories of mechanics to explain flight mechanisms according to their own fancy. And they of course insist upon it being printed."

For the first eight decades of the nineteenth century, the field of physics that describes the flows of gases, applied aerodynamics, effectively remained in its infancy. Englishman Sir George Cayley had published the fundamentals of a scientific approach to the problem of flight in a series of articles entitled *On Aerial Navigation* at the beginning of the century, discussing and emphasizing the importance of fixed wings for flying machines. This was an important step toward applied aerodynamics. Cayley had also suspected that curved surfaces might have properties that differ from currently accepted theory. But there was largely no follow-up to his work. Others had also noticed the characteristic curvature of bird wings, but no usable experimental database had been established.

Individuals studying bird flight often gave detailed descriptions of their observations, but typically did not present them within the conceptual framework of physics. Many of the theories of those described as *"aeronautical hotheads"* by Lilienthal, individuals who often claimed to have invented flying devices, contradicted the existing theoretical consensus. The highly developed theoretical mechanics of the nineteenth century explained flows from the perspective of hydrodynamics. This field relied on describing typical processes as stable, idealized flows. At the time, hydrodynamics couldn't account for unstable flows around asymmetric objects such as a bird's wing whose shape and attitude varied over time.

Consequently, the pioneers of applied aerodynamics in the nineteenth century were technicians whose investigations of the subject were grounded in practical applications. However, two other properties played an essential role in allowing Lilienthal to make a major contribution to the physical description of wings: first, he was able to spend two decades researching the fundamentals, enabling him to delve beyond the practical aspects of his subject. He developed measuring devices and methods to describe wings within a conceptual framework. He returned to practical applications once he was satisfied with his theoretical foundations and had published his results. He paid to have his book printed out of his own pocket, but it was worth it to him. The book ends with Chapter 41: *The design of a flying device*, containing a compilation of

> "*points [...] that the design of a flying device would have to observe if the experimental results published in this book are acknowledged and the views developed from them are correct.*"

And indeed, they were. Before two years had passed, the author himself had already put his points into practice by developing a marketable product that built upon the scientific foundations he had established earlier. But Lilienthal's third characteristic was most decisive of all: he was sufficiently athletically gifted to teach himself to fly with his own flying gear and demonstrate the accuracy of his theories with self-conducted experiments. The photographs of the *Flying Man* attracted international interest far beyond the small community of aviation pioneers.

But now we need to go back to the beginning. By studying kites, Otto had convinced himself that a kite tethered by a string could only move if either the air moves around it, or the kite is pulled forward through still air. From this, Otto concluded that an aircraft would only be able to stay in the air while being propelled. Otto's experiments with his small steam engine were the first logical step in this direction. However, the air resistance on the wings was still too high, so Otto's next task was to optimize the wings. But how? He needed to measure two things—the usable lift and the drag on the wing. To do this, Otto constructed a device that resembled similar inventions by others. Lord Cayley's *Whirling Arm* device in particular deserves to be mentioned here, which was used to measure the lift and drag of flat plates under inclined flows.

Lilienthal's device for investigating the air resistance and lift on flat and curved surfaces consisted of two very lightweight opposing arms that were mounted on the center of a rotating vertical spindle. This rests on the arm of a balance. If the weight of the moving parts changed during rotation, this

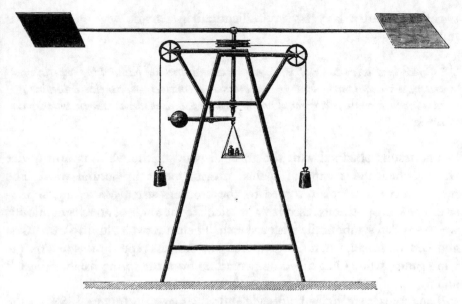

Fig. 1 Lilienthal's *Rundlaufapparat (Whirling arm Device or Aerodynamic Balance)*, a device for simultaneously measuring the lift and drag on surfaces as a function of the angle of attack and the speed. Illustration in Lilienthal's book, p. 61

could be measured on a spring scale. Test surfaces could be attached to the arms at any desired angle. A weighted rope was wound around the middle of the spindle, and the apparatus was wound up like a grandfather clock. When the weights were released, the spindle was set in motion. After a brief spool-up, the rotation became steady. An equilibrium was established between the torque of the falling weights and the air resistance of the test surfaces, and different weights yielded different flow velocities. Lilienthal built several such devices with diameters ranging up to seven meters, and recorded wind speeds of up to twelve meters per second. He carefully examined and accounted for any possible sources of measurement error, including the air resistance of the arms and friction from the machine itself. He was exceptionally patient and methodical, traits that made him exceptionally well suited to conduct experiments.

The series of measurements began with flat test surfaces and were then extended to highly diverse profile shapes. Measurements were first taken in a gymnasium, then on a wide treeless plain at the gates of the city, between Charlottenburg and Spandau. The results showed that slightly curved wing surfaces with a low angle of attack, as can be found in the shape of a bird's wing, yielded by far the best values, meaning the lowest air resistance and the most lift. Ludwig Prandtl, the most renowned aerodynamicist of the early twentieth century, later assessed Lilienthal's systematic way of working as follows:

> "*We see how every piece of knowledge was acquired by painstaking experimentation, with equipment built by the brothers themselves, nevertheless achieving a measurement quality that would not be surpassed until modern research with wind tunnels.*"

The results obtained with the artificial wind of the whirling arm device were verified and compared against measurements in natural wind. The measurement protocols recorded by the brother's sister, Marie, are impressive. The wing positions, used later to calculate the angle of attack graphically, were recorded symbolically with a sketch. There are two columns, titled *Otto* and *Gustav*. It seems that Gustav called out the wind speed measured on the wind gauge, while Otto called out a reading from the spring balance (Figs. 2 and 3).

The results were verified a decade and a half later, in August 1888, on the plain between Teltow, Zehlendorf, and Lichterfelde, near the *Royal Prussian*

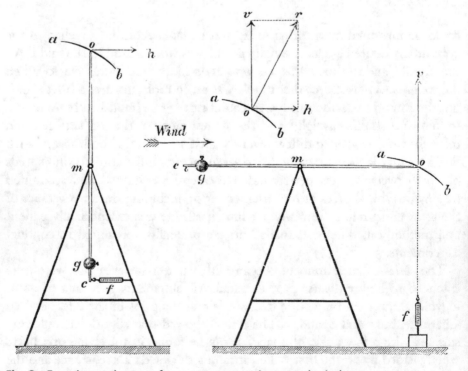

Fig. 2 Experimental set-up for measurements in natural wind

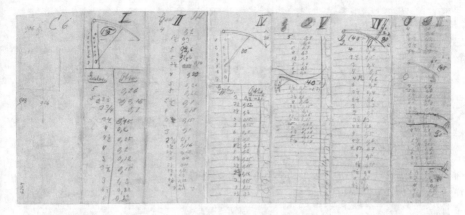

Fig. 3 Measurement protocol, 1874. © Deutsches Museum. All Rights Reserved

Main Cadet Institute, which became the Andrew barracks of the US Army after World War II. Despite much larger test surfaces, a larger circuit, and different building materials, the same values were found. To gain insight into the best materials, the test surfaces were made from thin pressed cardboard, solid wood, and brass sheets. The brass sheet, thickened at one edge with a wire, was found to be less favorable than the other materials. The final shape, which resembled a curved droplet, looks very much like a modern airfoil. This was also noticed by Lilienthal, who modelled the profile after bird wings, which have bone at the leading edge (Fig. 4).

> *"It even seemed as if this shape had particularly favorable air resistance properties, with significant lifting and little inhibiting resistances, especially when moving at very acute angles, but only if the front edge was thickened and not the rear edge."*

However, the differences measured by Lilienthal between the profile shapes shown in his Figs. 8, 9, 10, 11 and 12 in Chap. 8 were small, so he only used the parameter of curvature in his subsequent measurements (Fig. 4).

Attending the *Royal Trade Academy* in Berlin had given Lilienthal an excellent education in the technical sciences, which were still developing at the time. His working methods and style were influenced and shaped by one of the school's technical scientists, Ferdinand Jacob Redtenbacher. Redtenbacher believed that the laws of mechanics could be seen everywhere, and that theory, engineering, and technology were deeply intertwined. *"Wherever anything stirs, mechanics are at work; yet spirits are not moved by mechanics,"* he wrote.

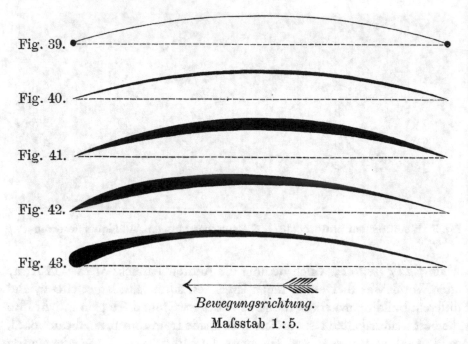

Fig. 39.

Fig. 40.

Fig. 41.

Fig. 42.

Fig. 43.

Bewegungsrichtung.
Maſsstab 1 : 5.

Fig. 4 Even with the same surface, curvature, and shape, differences in the material and production created variations in the airfoils, which were precisely compared against one another

Redtenbacher's philosophy was popularized by his students, who also taught at the *Royal Industrial Academy* in Berlin. The academy's director, Franz Reuleaux, structured his teaching according to certain fixed principles. Every week, there were 36 mandatory hours of theoretical lessons and practical work in the workshops. This resulted in studies similar to those of a technical university.

This period clearly had a formative effect on the young engineer that persisted throughout his entire life, in line with the best traditions of German engineering. His success as a mechanical engineer and aviation pioneer came from a fortuitous blend of excellent education, talent, and passion, combined with many experiences from his childhood and youth. Although Lilienthal agreed with his teachers on many things, his views diverged strongly in other regards. When Reuleaux offered him the position of an assistant, considered to be highly desirable, Lilienthal turned it down—he had already made other plans. Lilienthal was drawn to practical experiments, to applying what he had learned, to his own inventions—and of course to the invention of human flight. It is possible that the work of the so-called *Helmholtz Commission*, a body commissioned by the government to study aerodynamics, also contributed to his decision. Another researcher, Adolf Hörmann, conducted experiments with a rotating measuring device at the *Trade Academy*, and it seems likely that Lilienthal also participated. He cited the commission's work in his first lecture on aeronautics in 1873.

In Lilienthal's time, there were very few reliable theories of aeronautics, unlike mechanical engineering, which was already a relatively mature field. Lilienthal soon realized just how much the contemporary understandings of aviation diverged from one another, both in theory and in practice. Isaac Newton's classical formula for air resistance was inadequate since it only describes the resistance of a flat surface perpendicular to the wind. Otto's measurements showed that this model was clearly not suitable for describing the processes generating and affecting the lift of a wing:

"More precise knowledge of this air resistance is unfortunately limited to a few very simple applications. Air resistance is only truly understood in the case where a thin, flat plate moves through the air in the direction perpendicular to its surface."

In practice, there was a great need for a method of predicting the air or fluid resistance. These quantities play an integral role in calculating the trajectory of projectiles or building ship hulls. Understanding the force exerted by the air is also important in structural engineering. Alongside the classical values based on Newtonian mechanics, the technical manuals of the time listed empirical values that differed strongly. Knowledge of hydrodynamics was simply transposed in an attempt to describe aerodynamics.

There were no other theoretical tools available to the aviation pioneers of the nineteenth century, the enthusiasts who were actually experimenting in the field. In 1842, Englishman William Samuel Henson patented his design for an *Aerial Steam Carriage* that already featured many of the essential aspects of a modern aircraft, though his subsequent attempts to achieve manflight with his aircraft were destined for failure. Henson was far from alone in this regard. There is a long list of pioneers who failed in their attempts to fly: in 1883, Russian Alexander Fedorovich Mozhaysky made small jumps into the air with a gigantic machine powered by steam engines.

Sir Hiram Stevens Maxim, an Englishman of American descent, invested a fortune in a massive flying machine that boasted a 31.5-m wingspan and was powered by two steam engines with propellers. In 1894, the machine actually managed to lift off the ground once before the experiments were stopped.

In 1897, a flying machine built by Frenchman Clément Ader took off briefly, as did Langley's Aerodrome six years later, famously crashing in the Potomac River.

In addition to the lack of a theory to describe airflows, empirical aircraft designs faced a second problem—how can even an airworthy flying machine be controlled in flight, and how can stable flight be achieved? Lilienthal gave a convincing description of this problem based on his own experience in a lecture on November 6, 1894, to the *Berlin Architects' Association*. According to Lilienthal, the problem is that

"You can only learn to fly by practicing, but you can only practice flying without breaking your neck if you already understand it! This is precisely why the question of flight has not been solved to this day."

The practical failures of other would-be aviation pioneers and the inaccuracy and contradictions of the research they relied on had a great impact on Otto Lilienthal, much like his lifelong observations of birds in flight. In his lecture at the *Potsdam Trade Association* in 1873, Lilienthal spoke of being inspired by bird flight and "*listening to the birds'* art of flying". He continued:

"Today, we are used to attacking problems like this analytically, but the art of flying specifically is not susceptible to simply being invented like gunpowder."

This insight underpinned his general strategy. Not only did Lilienthal observe birds in flight, he also sketched them, wrote poems about them, and studied them with scales, a watch, and a tape measure. He delved in this subject both scientifically and emotionally and continued to do so for many years. His scientific investigations bear the hallmarks of an experienced practitioner and attentive observer:

"In general, nature seems to prefer having a smooth upper surface; at the very least, this is evident in the shape of the flight feathers, which are always perfectly smooth on top. [...] From this, we can conclude that the suction effect over soaring surfaces is more significant during flight than the pressure effect of the air on the underside of these surfaces."

What Lilienthal expresses here as a hypothesis is now considered common knowledge. His self-designed measuring devices were not like academic equipment with extremely fine mechanical precision. They were large and robust, delivering results that could be repeated and verified at any time, and which allowed the results to be tested and evaluated, and made it easier to identify and account for measurement errors and other anomalies.

Lilienthal's approach was always systematic. His lectures and his book *Birdflight as the Basis of Aviation* reveal how he worked and highlighted his logical process: he observed the birds, then measured their performance and conducted various precise studies while making detailed calculations. He determined the work done during each upstroke and downstroke of their wings, and the effects on each phase of flight, postulating the performance of the birds' muscles in the process. Lilienthal learned as much as he could about morphology and behavioral biology, and transferred his observations to technology and the laboratory. He imagined and built experimental devices that simulated the natural processes he'd observed. He did not need to, nor could he, rely on an existing body of knowledge, seeking instead to verify all of the basic principles with his own measurements before building on them. He studied the available botanical knowledge of winged plant seeds and their properties and diversity. He measured and observed different birds, and compared his observations of their structure, behavior, and performance with his own calculations.

From this careful study, Lilienthal concluded that there are no specific "*biological factors*" in the feasibility of flight. That is, the flight of birds must also obey the laws of mathematics and physics. This may seem simple in hindsight, but, to Lilienthal and those who followed, it provided a crucial and foundational proof that flight was possible.

"Nature will not do anything without a special purpose," he wrote at the time. "There truly is no law of nature preventing the solution of the problem of flight like some insurmountable barrier."

All of his equipment and devices were self-designed and self-built. And while his predecessors performed single experiments, Lilienthal conducted experiments in series, as he was taught in school. He stayed rigorously true to the scientific method, only trusting his conclusions when different experimental setups generated the same results. Lilienthal believed in science, not miracles. His goals for his experiments were clearly stated:

" [To] become more familiar with the peculiarities of air resistance phenomena and thereby stimulate further research into exploring the most important fundamental principles of flight technology."

In *Plate VIII* at the end of his book, Lilienthal gave a graphical adaptation of technical knowledge derived from the model provided by nature. The force diagrams of each profile are drawn onto a stork's wing. Among the 30 principles for building aircraft published by Lilienthal in chapter 41 of this book, the following of his conclusions are still valid today:

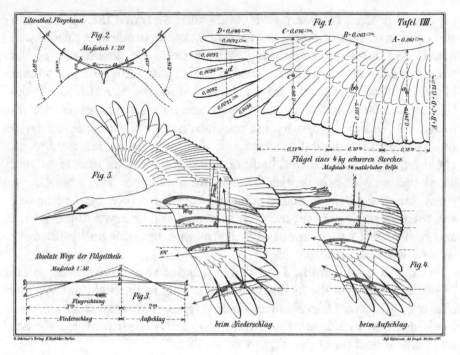

Fig. 5 One of Lilienthal's most famous charts: Plate VIII, the last plate published in the appendix of his book

13. *"The wings must show a curvature on the underside.*
16. *The ribs and stiffeners of the wings should be fitted near the leading edge.*
18. *The geometrical shape of the curvature should be a parabola, more curved toward the leading edge and getting straighter toward the back.*
19. *For larger areas the best shape of the curve would have to be established by experiment, and that shape, the pressures upon which for small angles of inclination are most nearly in the direction of movement, should be preferred."*

In 1882, VFL began publishing a journal, the first specialized publication about aviation in Germany. The journal published articles about topics such as balloons and their applications in science and the military. Aeronautical questions also gradually began to be discussed for the first time. In 1892, Lilienthal published an article in this journal, boldly stating the need for an engineering approach to the pursuit of flight. In the piece, titled *On Mechanics in the Service of Flight Technology*, Lilienthal chided those who were interested in the field, but lacked the knowledge to make meaningful progress.

"We often see publications from the hand of laypeople who have never truly studied the actual mechanics. Their descriptions and observations of flight processes are often highly stimulating and genuinely valuable. However, as soon as these writers turn to mechanics to draw conclusions toward solving the problem of flight, they venture beyond the boundaries of their knowledge and usually follow deeply erroneous paths. [...] Furthermore, it is useless to attempt to persuade these emotionally driven practitioners of mechanics otherwise, as their lack of scientific understanding undermines any attempt to present a convincing argument. [...] Unfortunately, common sense is not enough in mechanics. [...] Without a thorough understanding of mechanics, nothing can be achieved in aeronautics. Mechanics is only a dry field to those familiar to it from afar. Fortunately, to perform aeronautical calculations correctly, only the most elementary of mechanical concepts are required; however, these concepts must be mastered thoroughly to guard against error."

In the spring of 1886, the Lilienthal brothers became members of the VFL, which had been founded in Berlin in 1881. Within this association, they were a minority, representing an aviation movement that stood opposed to the far larger faction of supporters who were invested in airships. Otto Lilienthal continued to play an important role within the association until his death. He became its secretary, served as a member of the editorial team for the association's journal, and for many years also as a member of the technical commission that examined all submitted proposals and inventions. More than anything else, however, he distinguished himself through his regular lectures to the group. Before his famous book was published, he gave three lectures in autumn 1888 and spring 1889 on the power consumption of bird flight and what that meant for the feasibility of manned flight. The lectures gave a preview and summary of the technical contents of his book. The minutes of the general assembly show that Lilienthal received an unusual honor after his very first lecture:

"At the chairman's request, the association's interest in the enthralling experiments of Mr. Lilienthal is expressed by the audience rising to its feet."

When his book was published the following year, the subtitle read:

"a contribution toward a system of aviation, compiled from the results of numerous experiments made by O. and G. Lilienthal, edited by Otto Lilienthal, engineer and machine constructor in Berlin."

As the author, this book represented a turning point in Lilienthal's aeronautical work. Its publication meant that his preparations for realizing human flight were now complete. He considered that his book had established the necessary theoretical tools and had paved the way to begin transitioning to practical applications of his work—actual flight tests. This transition required a well-defined representation of the physics of wings. Thus, the book concludes with a chapter entitled *The Construction of Flying Apparatus* which summarizes the principles

> *"...from which the construction of flying apparatus would have to be evolved, when the experimental results act down in this volume are accepted as a basis of the design."*

This is Lilienthal's personal program for his future work. And indeed, it would achieve practical success in less than two years, leading to Lilienthal's reproducible gliding flights and later developments of aviation and aircraft that continue to this day. Lilienthal struggled to find a publisher for his book and was forced to bear the printing costs for the first edition of 1000 copies himself. R. Gaertner's publishing house only later took over the book's printing and distribution. Robert Franke's xylographic workshop alone charged Lilienthal 359.40 marks for the 59 woodcuts needed for printing. A statement from the publisher to Lilienthal in 1893 shows that Lilienthal earned 100 marks for the sale of 20 copies in a year, or 5 marks per copy. To put that in some perspective, an average skilled laborer in Germany at the time might make as much as 1000 marks per year (though, as noted, Lilienthal was an advocate for better pay for workers), while a supervisor would see about 1500.

The book is remarkable for both its content and style. Lilienthal was especially concerned with presenting an accurate scientific basis for the phenomenon of flying. But he did not want to limit his audience to readers who were already educated in physics. Instead of scaring off his readers with formulas, he sought to introduce mechanics very carefully.

Accordingly, the book begins with a watercolor of circling storks, accompanied by the following words:

> *"With each advent of spring, when the air is alive with innumerable happy creatures; when the storks on their arrival at their old northern resorts fold up the imposing flying apparatus which has carried them thousands of miles, lay back their heads and announce their arrival by joyously rattling their beaks; when the swallows have made their entry and hurry through our streets and pass our windows in sailing flight; when the lark appears as a dot in the ether and manifests its joy of existence by its song; then a certain desire takes possession of man. He longs to soar upward and to glide, free as the bird, over smiling fields, leafy woods and mirrorlike lakes, and so enjoy the varying landscape as fully as only a bird can do."*

KREISENDE STORCHFAMILIE.

Fig. 6 *Circling family of storks,* a separately bound watercolor by Otto Lilienthal at the start of the book

"Who is there who, at such times at least, does not deplore the inability of man to indulge in voluntary flight and to unfold wings as effectively as birds do, in order to give the highest expression to his desire for migration?"
"Are we still to be debarred from calling this art our own, and are we only to look up longingly to inferior creatures who describe their beautiful paths in the blue of the sky?"

In Chapter 38, Lilienthal returns to his teachers not just for physics, but also for poetry. The physics text is briefly interrupted by two poems, one short and one lengthy. These poems were the only parts missing from the English edition published twelve years later, which otherwise followed the original text in full. The English edition was a translation of the second German edition, published in 1910, to which Gustav Lilienthal had added two chapters, one at the beginning and one at the end, describing the developments after 1889 and the flights that actually took place. The first edition stated:

The impression could be given that the only reason for the creation of the stork was to awake in us the desire to fly, to act as a teacher to us in this art. One almost hears him crying out a warning:

"O, see the delights that are with us there,
When, circling, we are borne in the warm blue air.
Stretching out far beneath our gaze
Your splendid, sun-lit world so fine.
And over us the sky's canopy sublime,
Where only your eye with us plays.

Our plumage carries us, our broad domed wing
Borne up high by the wind's lifting
Flight is no trial for birds:
No beat of the wings disturbs the sublime
peace. O man, in the dust, when will you find this ease?
When will your foot be loosed from the earth?

And when evening arrives, and the air grows calm
We come again down in the golden balm
And from the lonely heights depart.
The beat of our wings carries us peaceful and light
Toward the village before the sun's flight
Then we want to be where you are.

So you see us carried in a low flight
Over the gardens in the red of twilight.
We return to our nest:
Soon we find sleep in our shrine,
Dreaming of wind and bright sunshine,
And giving our feather'd limbs rest.

If driven by yearnig, like we are in flight
To glide away in the realm of the hight,

Enjoying what flying can bring
Then see our wings and measure our powers
And study the lift, which force is ours,
Concluding the works of our wing.

Then seek that which carries us there
Whilest our pinions gently stroke the air
Whiles our flight goes on, ever untiring!
That which was bestowed by a gracious creation
May lead you then to the due realisation
And solve the enigma of flying.

O! Just apply the power of thought:
An eternal ban must not be wrought -
also you will be borne by the ether.
It cannot be your Creator's will
To doom you, the first He made, to earth until
Eternity, to refuse you flight for ever."

From Theory to Flying Machine

From 1891 onward, Lilienthal reported to the association every year about his attempts at practical flying, the progress he had made, and the difficulties he had yet to overcome. The lectures were published in the association journal, which made them accessible to the admittedly rather small circle of people interested in aeronautics. After 1893, illustrations were added to the articles, including some of Anschütz's famous photos, which likely played an essential role in making the topic more interesting to the broader audiences of popular magazines and daily newspapers. As well as publishing in the association's journal Lilienthal regularly wrote articles in the popular magazine *Prometheus*, which described him as a colleague when it published his obituary. The weekly magazine was first published in 1889, and the foreword of the first edition described itself as:

> "...driven by the notion that any educated person will have an interest in regularly learning about advances in the natural sciences and their applications. [...] Our highest concern was therefore to ensure that our staff have excellent knowledge of all subjects. Our magazine will focus on original articles, offering several of them to our readers with each new issue; where the subject requires it, we will explain the text with carefully prepared illustrations."

This enhanced Lilienthal's influence on the development of aviation even further. People ultimately began to speak of a "Lilienthalian" school of thought in aviation, whose ranks boasted notable pioneers such as Alois Wolfmüller in Munich, Raimund Nimführ in Austria, Percy Pilcher in Great

© The Author(s), under exclusive license to Springer Nature
Switzerland AG 2022
M. Raffel and B. Lukasch, *The Flying Man*, Springer Biographies,
https://doi.org/10.1007/978-3-030-95033-0_7

Britain, Ferdinand Ferber in France, and finally Octave Chanute, and the Wright brothers in the U.S. Like Lilienthal himself, they all began their practical flight attempts with simple gliders, finding much more success than others who were attempting to force their way into the air in large motor-driven aircraft without sufficient fundamental research and understanding.

Lilienthal maintained correspondence with numerous others who were experimenting with aviation, both in Germany and abroad, and many of them visited him at his flying hill in Lichterfelde. He never refused a flight demonstration, and never denied anyone a tour of the factory where his flying machines were built in a corner. Lilienthal deliberately sought to exchange opinions and information, while keeping a watchful eye out for potential buyers of his flying machines. He was deeply convinced that flying could not be invented by any single person, but that a multitude of parallel endeavors were needed to solve the problems, as had been the case with the development of the bicycle.

He considered the development of flying machines to safely practice gliding and soaring to be his most important contribution. This was something that he repeatedly emphasized in his final years. He also imagined that the future would hold artificially built take-off points where athletically inclined youths could compete and measure their strength against one another with gliding contests. His vision of a glider airfield was realized some 25 years later on the Wasserkuppe mountain in Germany.

In the autumn of 1895, Lilienthal presented the following figures to aviation pioneer Octave Chanute in Chicago: his glider would cost around 100 marks each to produce in large quantities. Since he had already successfully sold several aircraft at a price of 500 marks, series production would yield good profits that could adequately compensate the years of effort and risk, as well as the financial investment.

After Lilienthal's death, Chanute wrote:

> "*This deplorable accident has robbed us of the man who has contributed most toward ensuring that humankind will one day likely be capable of flight; he was the first man of our times to attempt to imitate bird flight with a device adapted to human height, and he was so gifted in every respect that, had he survived, he would undoubtedly have succeeded.*"

As he rigorously studied the flight of birds and conducted his associated experiments and research between 1866 and 1889, some lengthy interruptions notwithstanding, Lilienthal established a key distinction early on. He realized that there were significant differences between the upward flight of birds, which required tremendous physical exertion, normal horizontal flight

Fig. 1 Reports on Lilienthal's flights in the *Aeronautics Journal* published by Octave Chanute

with moderate flapping, and the seemingly effortless gliding, circling flights that birds could make while still gaining altitude.

The belief that the secret to bird flight relies on flapping wings has a long tradition. Even the Lilienthal brothers' earliest experiments, performed to the standards of scientific rigor and described in Otto's book, involved a large wing flapping apparatus that had been scaled up to human size. They built a pair of wings with a wingspan of six meters, designed to be operated by the strong muscles in the legs. But even the greatest of exertions could only produce a lift of about 40 kg. The brothers needed a measuring device to imitate wing flapping in laboratory conditions without the exertion of lifting their bodies to solve this mystery. As before, the propulsion needed to be precisely measurable and generated by falling weights (Fig. 2).

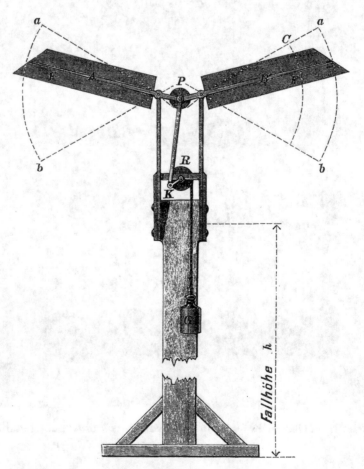

Fig. 2 Measuring device for studying wing flapping in the laboratory

Are birds really that much stronger than humans? Or was there something wrong with their wing flapping device? What was different about the flight of birds? Many birds also find it much more difficult to take off than to fly. The young Lilienthal brothers had already remarked that the storks they were observing could not take off down wind. If the brothers snuck up on them from the direction of the wind, the storks would even run toward them, flapping their wings violently, in order to take off. Once they were in the air, they could fly large distances without flapping. Was the wind the secret to flying?

To derive a valid formula that would describe the flapping wing motion as accurately as possible, the force required to flap the wings was measured, followed by experiments with free-flying small flapping-wing models in the early 1870s. Lilienthal understood that the number and amplitude of the wing beats would be decisive, as would the shape of the wings.

He had likely already experimented with flapping wing models as a student, since in a letter from deployment in March 1871 he asked his brother in Berlin what progress the latter was making with his models. He added:

"The device with the four spiral springs must have broken in two, because it was made much too delicately. I would like to expand the experiments in another direction, namely to investigate how the lifting effect changes when the device is simultaneously moved forward while the wings are flapped."

He had already realized that forward motion was at least as important to flying as the flapping of wings. He tested several new models of various sizes with wings operated by springs or even steam power until 1873. The models flew at different speeds both indoors and outdoors. Later, Lilienthal would describe his attempts with a steam engine as unsuccessful, but his business was founded entirely on the basis of small, light steam engines. They provided a livelihood for Lilienthal and were therefore indirectly essential for his aircraft construction efforts.

Conceptually distinguishing between these two phenomena, the lift generated by the shape of the wing and the combined lift and thrust generated by the flapping of the wing, was a decisive step forward. The new plan, which would ultimately be successful, was to research the generation of lift with suitable wing shapes in order to realize gliding flight. Lilienthal compensated his lack of a propulsion system and utilized the forces of nature by starting from elevated points, first a pedestal in his yard, later the wall of a gravel pit, then a shed on a small hill near his house, and finally in the Rhinower Mountains 60 miles away. After 1894, the artificial *Fliegeberg* in his hometown of Lichterfelde became his permanent airfield.

"Flying means: taking off into the air with a flying machine. We can't do that! Flying also means: travelling through the air from one mountain peak to another mountain peak of the same height. We can't do that either! Finally, flying means: lowering oneself through the air from the peak of a hill to the bottom of a valley. This is something we can do."

Lilienthal outlined his approach with these words in a lecture in 1894. He had learned to fly down a valley from a height. That was the starting point for all further attempts. But Lilienthal's preoccupation with wing flapping was not over once he had mastered gliding flight. His next goal was not to immediately achieve take-off by wing flapping, but to generate additional lift and extend his gliding flights.

At the time, Lilienthal thought that birds were able to soar because of some constant lifting wind. While winds like these do exist, especially in areas of rising terrain, he didn't consider the concept of thermals, which birds—and glider pilots—rely on extensively to gain altitude while circling. He was close, though, correctly surmising that the sun was the primary cause of air movement, and even wrote an essay titled *The Flight of Birds and Man Through the Warmth of the Sun*, but it doesn't mention thermals. Lilienthal based his initial theories on the idea that, because of the disruptive effects of ground features, wind speed tends to increase with altitude. He built an elaborate 10-m-high measuring device, and found that wind blows diagonally upward at an angle of about 3.5 degrees from the horizon. There's no reason to doubt his measurements, but Lilienthal's theory does not correctly explain soaring as we understand it today (Figs. 3 and 4).

Lilienthal was the first aerodynamic researcher to graphically represent his measurements as polar diagrams in his book. For each value of the inclination, the lift was plotted along the vertical axis, and drag along the horizontal axis. In the book's appendix, the polar values of the air resistance on flat and curved surfaces with different amounts of curvature are plotted in several plates. For a flat surface, the curve is approximately a semicircle. For the curved surfaces that Lilienthal investigated, the curve is strongly shifted toward the top left: these wings generate high lift and low drag (Fig. 4).

If Lilienthal is considered the father of human flight today, this polar diagram is a visible example of his legacy. Even today, aerodynamics experts and aircraft manufacturers study *Lilienthal Polars* as a method of representation to describe flow profiles.

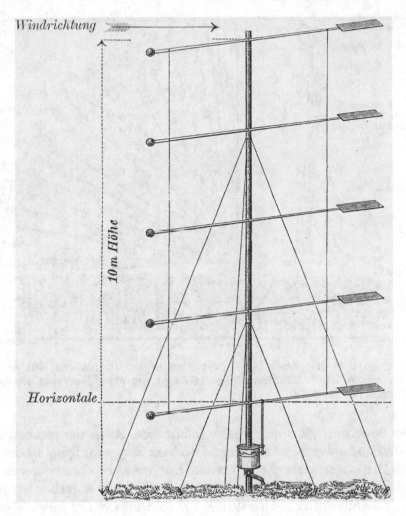

Fig. 3 Equipment used to measure and record the fluctuations in the wind's direction above the ground

After Otto finished his book, he set to work developing his first pairs of wings modelled after the principles he'd described in the last chapter. One of these principles states that the construction of "usable flight apparatuses" does not necessarily imply powered flight. This is followed by the observation that human strength alone is not sufficient to keep a flying machine in the air without the help of natural wind. According to Lilienthal's measurements, mastering winds with speeds of 10 m per second would be enough to achieve sustained gliding flight. Flight training, including in strong winds, always remained part of Lilienthal's program and is no doubt responsible for his fatal accident.

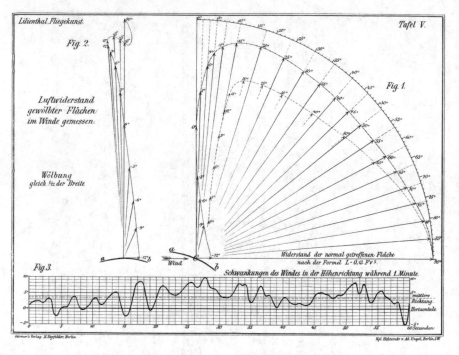

Fig. 4 Plate V in Lilienthal's book: evaluation of the fluctuations and average increase in the wind's direction as the cause of the more favorable lift values measured in natural wind

The book then gives quantitative information about the required size, cross-section, and shape of the wings. Two basic shapes for flying models are proposed, stating that the choice between them would be a matter of practical experience. One of these shapes is based on birds of prey with their feathered wing tips, while the second has slender tapered wing tips like gulls and other sea birds. Both of these shapes are reflected in Lilienthal's aircraft designs. The latter was already being used as a standardized model for the measuring surfaces in his experiments to determine the forces of the air (Fig. 5).

At first, Lilienthal did not consider any control or stabilization elements, like the tail fin. His twelfth principle states that they would have "subordinate importance for the lifting effect". The last ten principles discuss the implementation of wing flapping in an artificial wing. The earliest designs of his flying machines to carry a person have only survived in the form of sketches. No photographs were taken of his gliders until 1891. The machines built in 1891 and 1892 followed his proposal of a "pointed shape like sea birds". The later transition to a wider shape was not motivated by aerodynamic reasons but was made because of the foldable design of the bat-like wings introduced

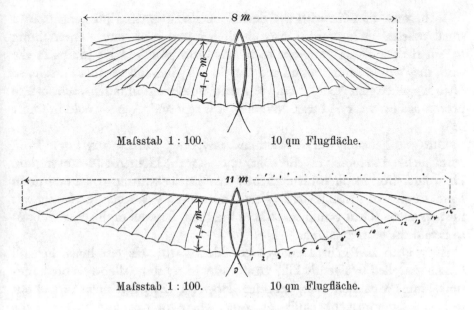

Fig. 5 First drafts for a flying machine, based on different species of birds

in 1893. This resulted in a completely different structural design. The feathered wing tips were incorporated into two different engine-powered wing flapping devices that were tested from 1893 on.

The wings had a recess in the middle for Lilienthal to stand, holding the flying machine with his hands and arms. As he transitioned from theory to practice, he needed to learn to use his wings, just like the young storks that he observed in his yard. The storks would spread their wings and make small hops and jumps before finally taking flight for the first time. Lilienthal decided to learn to fly in exactly the same way. His approach was later described as "the Lilienthal school" with the motto: "from step to jump, from jump to flying". In a lecture held in 1891, he said:

> "I am slowly exhausting the materials for my study room and elementary experiments. Now I must step out into nature, into the air and wind, armed with the knowledge I have acquired, and test the theories that I have developed with the flying machines constructed from them, among the very elements for which they were designed and built."

> "The discovery of the laws of active flight and the derivation of relevant theories has been the first part of the aeronautical endeavor. Now the construction and execution of flying machines and the organization of practical flight experiments must follow as the second chapter of invention in flight technology."

To become familiar with the device, he first practiced jumping from a small pedestal in his yard. Otto's children found it extremely entertaining to watch their father's attempt to fly. As Otto practiced his jumps in the yard, they would romp around and play in the wind and weather outside, so Otto suggested that Agnes should sew their clothes out of burlap sackcloth to protect against wear and tear. Naturally, it is reported, Agnes would not hear of it.

Otto was generally very relaxed and easygoing. On the way home from work, he liked to loosen his stiff collar and tie at the Lichterfelde train station. He would stroll through Lichterfelde's villa district with his collar and tie in hand, singing or whistling as he went. In the evenings, when the family was hosting guests, Otto would sometimes bid them farewell on his French horn to herald the journey home.

After Otto had gained enough confidence with his test hops, he and Gustav travelled to a gently hilly area nearby to try some slightly more ambitious jumps with their flying machine. However, this experiment turned out to be a disappointment. Unlike at home, where the trees and bushes in the yard provided some cover, there was no shelter from the wind out in the open, something that Otto had not considered. Even the weakest air currents made it impossible to hold up the wings, let alone attempt to steer them. Gustav later recounted how they were simply knocked over when trying to hold up the wings. After this experiment, Gustav no longer wished to actively participate in Otto's tests.

> *"When one stands in the wind with these wings, we can report from our own experience that one feels a lifting effect from the wind stronger than one would ever have imagined. Without training beforehand, human strength is simply not enough to operate such wings in the wind. The first outcome is therefore that we must carry our carefully calculated and delicately built device back home, smashed to pieces by the very first gust of wind."*

That was how Otto described his experience. He had not expected difficulties of the magnitude he was now facing. With the publication of his book, he thought that all the secret of flight had been solved and the foundations had been sufficiently developed.

An experimental protocol that Otto documented in 1890 described the introduction of a component that remains essential to this day—one that cannot be found on a bird. With an area of multiple square meters, Otto's wings had proven all but impossible to control, even in winds at lesser velocities than he'd stated as necessary in his book. His carefully planned experiments and meticulous and methodical research had become a challenging athletic exercise, requiring an almost artistic skill that he hadn't predicted.

Lilienthal wrote:

"Only standing exercises were possible with the flying machine. [...] If a wing ever rose too high, it was impossible to push it back down, and you were forced to give in to the wind to prevent the wings from breaking. The whole system then turned around, so that the wind was blowing into the wings from behind, flipping the apparatus over, and leaving you standing on your head."

Running against the wind was impossible because there was no way to maintain balance.

In his notes, Lilienthal proposed a potential solution:

"It seemed that it might be possible to automatically retain balance if a vertical tail were used to consistently align the device against the wind."

We do not know why or for whom Lilienthal was documenting the results of his experiment in such detail, though we are grateful that he did. For the next few attempts, he added a vertical tail fin to the device that acted like a weathervane, independently aligning the glider against the wind. Later, he added a horizontal surface to create a cross-shaped tail unit with both a horizontal and a vertical fin. This fundamental shape remains a core element of aircraft designs to this day.

Lilienthal did not find learning to use his wings as easy as the storks in his garden. He was confronted by questions that had not been addressed in his book. The book's content was limited to investigating the lift generated by wings; it can be seen as the birth of the aerodynamics of wings as a field. In the closing words of his book, Lilienthal had written:

"Birds fly because they work the air around them in a suitable manner with suitably shaped wings."

But how can you ensure that the lifting force is always acting on the correct part of the aircraft? On the ground, you can tip the glider forward or to one side. But an aircraft in flight can move in three spatial directions and rotate around three axes. How can you ensure that your flight is stable and controllable? This had not yet been recognized as a problem and was not discussed in the book.

For Lilienthal, learning to fly was a sport, a mission that required training and body control. Otto trained regularly, gradually figuring out that a wind speed of around five meters per second would be ideal for jumping into the wind. At this speed, the device became noticeably lighter, and he was able to perform longer jumps after a short run-up that allowed him to lift his feet off the ground. The few, typically low sand hills around his practice area, mostly just a few meters high, were no longer enough. The cadet school in Lichterfelde was therefore replaced by a hilly area west of Potsdam as a flight training ground. There, Lilienthal began a new series of experiments that marked the dawn of human flight.

1891—The Leap into a New Century

Derwitz is a small farming village with fewer than a one hundred inhabitants located west of Potsdam on the Magdeburg railway, between Werder and Gross Kreutz. Its most noteworthy feature is the old village church, a fortified medieval structure like many of the churches around Brandenburg; thick-walled, built from fieldstone and granite, with narrow slotted windows. The square tower and spire give a solid impression. But the altar side to the east features an elegant brick gabled building with Gothic arches and crenelations. The year in which the church was built is not known, though one of the old weathervanes was dated to 1630. At this point, Derwitz is thought to have been a settlement of shepherds for the nearby Lehnin Abbey. In this rural environment, about 25 miles from Berlin, Lilienthal performed his first true flights in 1891 (Fig. 1).

In the 1780s, Carl Otto Bournot, the descendent of a Huguenot family and the uncle of Gustav's wife Anna Rothe, served as pastor here. This connection is likely how Lilienthal knew about a hill that rises over the village to the north, in front of the railway. Officially, the hill is called the *Spitzer Berg* (sharp hill), but it was often shortened to *Spitzberg* or simply called the *Windmühlenberg* (*windmill hill*). Its pointed shape was exactly what Otto wanted, and the windmill itself played an important role in the history of flight, as the miller Hermann Schwach was a key assistant for Lilienthal's first flying attempts. Lilienthal stored his flying machine at the miller's homestead, which still exists to this day, after laboriously transporting it by train to the Gross Kreutz station.

M. Raffel and B. Lukasch, *The Flying Man*, Springer Biographies, https://doi.org/10.1007/978-3-030-95033-0_8

Fig. 1 The first photo of a man flying, taken by the meteorologist Carl Kassner, 1891. © Otto-Lilienthal-Museum.

The small hill was one of the sand dunes that abruptly towered over the plain here and there in the otherwise very flat Havelland region of Brandenburg. Lilienthal did not practice on the hill itself, but from a gravel pit facing the railway line, where sand had been excavated during the construction of the railway and was therefore slightly steeper. From the outermost point, he could jump west, north, and east, depending on the direction of the wind.

Much of the original slope was removed after the turn of the century to build the Charlottenhof and Wildpark train stations near Potsdam. For this

Fig. 2 Memorial for Otto Lilienthal on the *Windmühlenberg* between Derwitz and Krielow: Wilfried Statt, 1991. *Photo* Nitsch. © Otto-Lilienthal-Museum. All Rights Reserved

purpose, a wide swath was carved through the hill from the south. While the exact launching spot no longer exists today, a centennial memorial was erected in Derwitz about halfway between the original location of Lilienthal's flights and Hermann Schwach's homestead (Fig. 2).

In the summer of 1891, Lilienthal brought his flying machine to Derwitz by train. Schwach picked him up from the Gross Kreutz train station by horse cart. They had to drive for around two or three miles back along the railway to Derwitz, passing through the village of Krielow, before moving the flying machine into the Miller's barn (Fig. 3).

Schwach's house and barn were about 100 m (appr. 100 yards) below the mill, on the western slope of the *Spitzberg*. The glider, the one Lilienthal had used to practice jumping in his yard, had a wingspan of 7.5 m (8,2 yd), and a chord (the distance from the leading edge to the trailing edge of the wing) of 2 m at the widest point. The original wing area was 10 m² (100 sq.ft), but that had shrunk to 8 m² after multiple repairs and changes. Each wing had two spars made from strong willow branches, extending from the base to the tip. On these spars, 14 weaker rods were arranged as ribs running parallel to the direction of flight. The ribs were bowed, giving the surface the desired curvature. The wing was covered with shirting, a tightly woven cotton fabric, with a lacquer coating. The glider weighed in at 18 kg (40 lb.), and included

Fig. 3 The mill belonging to Schwach, assistant and eyewitness to the first ever flights made by man. © Otto-Lilienthal-Museum.

both the vertical and horizontal stabilizers that had been added during the Derwitz experiments (Fig. 4).

For the first time, the *Derwitz Apparatus* featured the crossbar or cross frame present in all of Lilienthal's later gliders. Two crossed square timbers connected the two wing halves together in the middle. This created a V-position for the wings, mimicking the natural dihedral of birds that plays a key role in flight stability. The spar cross was also fitted with two cuffs, in which Lilienthal placed his forearms to pick up the glider. With his hands, he gripped the ends of the front spars protruding from the wings in a underhand grip position. This fixed his forearms firmly into position, while leaving his legs and upper body free to run, land, and steer the glider in flight by shifting his weight. On the ground, the glider could be carried in the same position.

Fig. 4 Derwitz 1891—the beginning: The wing spars on both sides of the flying device are made from thick willow rods. There are also 14 thin, curved rods arranged as ribs in the direction of flight. The cross frame or spar that characterizes all of Lilienthal's gliders is made from pine wood. © Otto-Lilienthal-Museum. All Rights Reserved

The glider was tail heavy, but a good headwind helped shift the balance by lifting the tail. In flight, Lilienthal hung freely by his forearms, and could let go and fall away from the glider if he lost control. This "cockpit" built by the world's first aviator, which seems so unusual to us today, changed little over the course of Lilienthal's many designs. Today, we'd call it a hang glider, a type of aircraft that experienced a strong resurgence in the 1960s and '70s and remains a popular vehicle for sport flying around the world.

Lilienthal occasionally gave names to some but not all his many different flying machines. Sometimes he named them after the location of his flights, like the *Derwitz Apparatus* or the *Stölln Model*. Other times, he named the design drawings after the names of his customers, such as the *Lambert Modell* or the *Seiler's Apparatus*. In a few cases, a descriptive name is given— in addition to the *Normalsegelapparat*, he used terms like *Large Biplane* or *Small Flapping Wing Apparatus*. For the models that Lilienthal left unnamed, later generations have filled in the gaps in a similar style, such as the *Südende Apparatus* or the *Maihöhe-Rhinow Apparatus*.

Fig. 5 Replica of the *Derwitz Apparatus* in the Otto-Lilienthal-Museum. The cross frame and the ends of the willow spars that serve as handles are clearly visible. © Otto-Lilienthal-Museum. All Rights Reserved

The wing curvature of the Derwitz glider was about 1/10 of the wing chord. Lilienthal chose this ratio because he expected that the forces experienced during flight pushed the curvature down to 1/12 or even lower. But the wings were stiff enough that the curvature did not change much.

When Lilienthal gave a detailed report on the Derwitz experiments to the VFL in November 1891, he passed around a few photographs that showed him in the air. These photographs were taken by Kassner, Lilienthal's meteorologist friend from Berlin, who appears to have accompanied him to Derwitz on two separate occasions. Kassner, the first person to take photographs of Lilienthal, had joined the Prussian Meteorological Institute in 1890, whose most prominent members also played a leading role in the VFL, which is likely how he became acquainted with Lilienthal. Kassner joined the association himself in January 1892, the same year he earned his doctorate in meteorology. Later, he participated in several scientific balloon flights organized by the association, experimenting with new meteorological instruments and making a name for himself by capturing a series of excellent photographs of clouds.

Kassner's photographs were the world's first photographs of flight. After showing prints at the *International Airship Exhibition* in 1909 in Frankfurt am Main, he donated the original slides to the *Deutsches Museum*, unfortunately however, they are among the losses of WWII. Since he was familiar with all of Lilienthal's practice areas from repeated visits, except

Fig. 6 Otto Lilienthal (center) with assistants at the wall that he used as an airfield in Derwitz. *Photo* Kassner, 1891. © Otto-Lilienthal-Museum. All Rights Reserved

Fig. 7 Sunday excursion with flight on the program: Otto Lilienthal with his wife, four children, and friends, in front of the miller Schwach's property. *Photo* Kassner, September 27, 1891. © Otto-Lilienthal-Museum. All Rights Reserved

for the Rhinower Mountains, Kassner later played a key role in providing information about their precise locations.

The first time that Kassner accompanied Lilienthal was likely at the end of August, when Lilienthal was confident that he could show Kassner flights to be photographed. The date of the second photography session is known—Sunday, September 27. The flights took place during a family outing. The Schwach family welcomed Lilienthal's family with their four children, his assistant Hugo Eulitz from Lilienthal's factory, and others, probably including actresses from the *Ostend Theater*, as their guests. Lilienthal made the following report to the association:

"Almost every Sunday, and even during the week if I had time, I was at the practice area between Gross Kreutz and Werder to practice gliding against the wind at the hills thousands of times. A technician from my machine factory, Mr Hugo Eulitz, and I took turns to glide down from the hill and immediately carry the apparatus back up the slope, while the other was resting to begin the next jump without delay. This ensured that we could fully take advantage of our practice time. In this way, Mr Eulitz and I developed the ability to glide down the gentle slope under moderate winds and land safely at the foot of the hill."

Hugo Eulitz was one of the cousins of Agnes Lilienthal. He trained as a locksmith and joined the factory as a technician in 1889. During the Derwitz experiments, he was 21 years old, while Lilienthal himself was 43. He was also present at the experiments in Südende the following year, but after that he was no longer mentioned as a participant by Lilienthal. Gustav Lilienthal was not involved in the practical flight experiments at that time.

During the experiments in Derwitz, Lilienthal was able to glide about 20–25 m through the air from a jumping height of 5 to 6 m, both in still air and under winds of different strengths, as Lilienthal was keen to emphasize. The only difference was the duration of the flights: the stronger the wind, the longer the glider could remain in the air.

In Derwitz, Lilienthal also had to learn about a new phenomenon; if the wind suddenly picked up mid-flight, the machine would come almost to a standstill, relative to the ground. He reported being repeatedly lifted several meters into the air by sudden gusts of wind, only avoiding serious injury by quickly allowing himself to fall out of the glider as it was swept up. Any sprains suffered by his arms or feet during the Derwitz experiments healed in just a few weeks, he added.

Fig. 8 Hugo Eulitz, who also flew the glider, is in this photograph. At the time, Eulitz was 21 years old, working as a technician in Lilienthal's factory. He was the only assistant who also participated in flights. © Otto-Lilienthal-Museum. All Rights Reserved

The miller Schwach later spoke about an unfortunate flight on which Lilienthal sprained his foot:

On the way to Gross Kreuz, I spoke to him reproachfully. *"Mr. Lilienthal, you really don't need to be risking your life like that. You have a family."* A train was passing by as he replied to me. *"Mr. Schwach, do you imagine that, when the first railway was built, it ran as safely and smoothly as it does today? It will be the same for flying."*

Most aviation historians consider the gliding flights in Derwitz as the beginning of human flight. The French aviation pioneer Ferdinand Ferber, who enthusiastically admired Lilienthal, was the first to say so. In a lecture in 1898, Ferber said:

"I consider the day of 1891 on which Lilienthal first measured a flight of 15 meters in the air to be the moment when mankind learned to fly."

Ferber repeated this view several other times in his writings. Like Lilienthal, he began his gliding flights in 1898. In 1902, he travelled to Berlin and

Fig. 9 Lilienthal before landing in Derwitz. Before he touches down, he lifts the leading edge of the flying machine, increasing the angle of attack and reducing speed. The same behavior can be observed with every crow that sets down, he said in front of the VFL in Berlin. *Photo* Kassner, September 27, 1891. © Otto-Lilienthal-Museum.

laid a wreath at Lilienthal's grave in Lichterfelde. The ribbon bore the inscription: *Le Capitaine Ferber a son maître Otto Lilienthal* (*From Captain Ferber to his Master Otto Lilienthal*). Ferber himself died in a fatal crash in 1909.

Lilienthal concluded his report on the Derwitz flights with the words:

> "*The exercise area at this location did not allow me to fly long distances from greater heights; therefore, I feel compelled to find another location to continue these experiments, so that I may practice jumping from even greater heights to fly even greater distances. Nevertheless, these attempts have given me the conviction that diagonally downward gliding can be achieved with a very simple apparatus and may be practiced safely from any height.*"

Lilienthal's flights at this location successfully carried him over distances of up to 25 m. Having to travel to Derwitz simply to conduct a few short practice flights was of course very impractical, so he resumed his search for a more suitable and more convenient flying location. In 1892, Lilienthal found a new practice area on the *Rauhe Berge* (*rough mountains*) in Südende, at the edge of the Steglitz district—a hilly, sandy location not far from Lichterfelde. It was

Fig. 10 Flight in a Sunday suit, not his usual aviator's costume. *Photo* Kassner, September 27, 1891. © Otto-Lilienthal-Museum.

Fig. 11 Twelve photos by Kassner showing Lilienthal in flight have survived from 1891. © Otto-Lilienthal-Museum.

Fig. 12 Lilienthal with his new glider in 1892 in Südende. His brother Gustav is standing next to the left wing, and Eulitz is on the right. Lilienthal is resting his hands on the device's two-pronged clamping support. The horizontal tail surface is not yet built into a cross together with the vertical surface, even in this glider from 1892, but is instead installed in front of it. *Photo* Kassner, August 7, 1892. © Otto-Lilienthal-Museum. All Rights Reserved

only around 20 min' walk from his house on the Boothstrasse. The new location was another gravel pit with an edge that Otto could jump down from. After a short run-up, there was a drop of several meters down to the bottom of the pit. Leaving the ground behind so abruptly was of course unfamiliar and not entirely without risk. Obstacles like cart tracks made training even more unsafe (Figs. 12 and 13).

The *Rauhe Berge* are foothills of the Teltow plateau, which borders the Berlin valley from the south, much like Barnim does to the north. The surface of the Teltow is mostly loamy sand, with some purely sandy areas around the foothills containing gravel deposits. The highest point of the *Rauhe Berge* is east of Südende, at an elevation of slightly more than 60 m. Lilienthal's practice area was to the west of this point, on a hill on the other side of the Anhalt Railway called the *Steglitzer Fichtenberg* on some old maps. But to the farmers who lived in Steglitz, the hill was simply called the *Rauhe Berge*.

The *Rauhe Berge* between Steglitz and Südende was not used for agriculture. Its wide sandy expanses were covered with grass, flowering weeds, and

Fig. 13 Lilienthal in his usual aviator's costume above his airfield. *Photo* Kassner, August 7, 1892. © Otto-Lilienthal-Museum. All Rights Reserved

veins of mud. The farmers had excavated a large amount of sand, creating a high wall that Lilienthal could use for his jumps. To the east, the villa settlement of Südende was just beginning to emerge. In his 1892 annual report, Lilienthal wrote about his experiments in Südende:

"*Having learned a lot from my experiences the previous year, this year I attempted to perform gliding flights with wings of up to 16 m^2 surface area. My largest device weighed 24 kg, which together with my own weight of 80 kg gives a total load of 104 kg. A wind speed of 10 meters per second should therefore suffice to keep me in* *the air with the device. But I have been careful not to expose myself to a wind this strong with such a large device, using smaller wing surfaces for my gliding exercises this year with air currents of this strength."*

A major disadvantage of the Südende wall was that it only allowed jumps to be performed to the west. Over the summer, there were many southerly and easterly winds, meaning that Lilienthal was periodically grounded. Nevertheless, he reported having found several opportunities to try his large device in the wind, and with practice he succeeded in safely gliding down from a height of 10 m.

"Spectators on the edge of the cliff who wished to observe my endeavors, initially declaring me to be some daredevil, assured me after the first few flights that the motion through the air gave an impression of complete safety, and that it was a beautiful and satisfying sight to see the large apparatus glide so calmly."

The strongest winds that Lilienthal had dared to face with his large glider had a speed of 7 m per second, according to his own estimate. The first part of the gliding flight after the jump was almost horizontal, then the line of flight descended until landing. In the most favorable cases, the flight distance was eight times the jump height, i.e., around 80 m. This was a considerable improvement over the results achieved in Derwitz, not only in terms of distance, but also in terms of what we now call the glide ratio, clearly established as 8:1.

Landing with the large glider, which Kassner once again captured in photographs, could always be done very softly on the sandy ground. To demonstrate this clearly, Lilienthal would often touch down with only one foot and remained posed that way for a while without so much as swaying.

Lilienthal wrote about the flying devices in his annual report:

"The wings that I used this year were a little less curved that the wings from last year. The benefits were unmistakable in strong winds. Once again, each device was fitted with a vertical and horizontal tail. Without these tails, the exercises in the wind would have been quite impossible."

If his words are to be taken literally, Lilienthal must have practiced with devices of different sizes in Südende. He explicitly talked about the new, large glider with a surface area of 16 m^2, which is the one that Kassner photographed. The pictures from 1891 were taken over two days, but those from 1892 were probably taken on a single day. Much of our knowledge

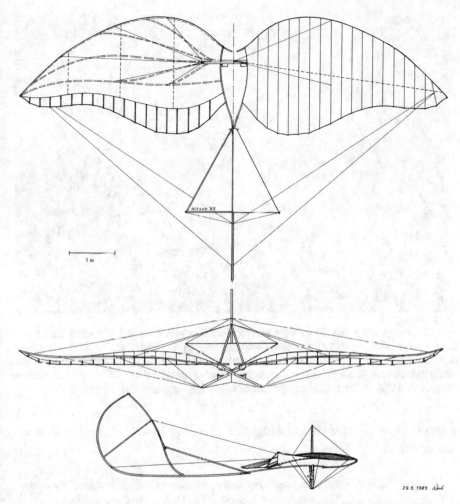

Fig. 14 *Soaring apparatus built around falsework* by Lilienthal in 1892, with a wing area of 16 m² and raised wing tips. The inscription states that the flying machine was built around "falsework", i.e. a prefabricated template. Each wing has two primary spars and four auxiliary spars, as well as 18 curved ribs arranged in the direction of flight. Eighty percent of the surface was covered on both sides of the wings. © Otto-Lilienthal-Museum. All Rights Reserved

of Lilienthal's flying machines comes from these photographs, as well as the large number of design drawings and sketches that have survived. For many of them, we no longer know whether they were unrealized designs or flying machines that were actually tested and forgotten. It seems likely that others existed, as Lilienthal is explicitly referring to the "large" glider. Furthermore, in a letter from Lilienthal on December 8, 1892, to a Sigmund Strauss in Vienna, Lilienthal recommends beginning by building a very simple glider,

Fig. 15 Jump from the 10-m-high wall in in Südende. Like the previous flight exercises in Derwitz, spectators often came to watch in Südende. Lilienthal's gliding flights gave them the impression of being completely safe, he reported. The best results achieved in Südende were flights over a distance of 80 m. *Photo* Kassner, August 7, 1892. © Otto-Lilienthal-Museum.

around 10 m² in size with a wingspan of 8 m. He writes along with a very simple sketch:

> *"I happily comply with the request for more detailed information about the apparatus I use for gliding exercises. I intend to publish a larger publication on a practical construction of the apparatus at some later date. However, because I still want to implement and test many new ideas in the near future, I cannot yet begin with these publications.*
>
> *I would simply like to recommend that you begin by building an apparatus that is as simple as possible, about 10 m² in size, such as the one shown in the sketch, which is has a wingspan of 8 m and a width of 2 m. The strong rods consist of 25–35-mm-thick willow shoots, which taper toward tips that are approx. 15 mm thick. If not available as a single piece, they must be joined together. The thin cross rods are 10–14 mm thick. The dotted lines are galvanized iron wire, approx. 2 mm thick. This wire should be doubled from c-d."*

Fig. 16 This picture shows a dangerous flight situation. Lilienthal needs to shift his body weight far backward to increase the angle of attack. But his arms are stretched out too far, and from here he is no longer able to bring his body back to the pilot position by bending his arms. Lilienthal later counteracted this danger by adding upper arm pads that secured the arms in a bent position and prevented the body from sagging. *Photo* Kassner, August 7, 1892. © Otto-Lilienthal-Museum. All Rights Reserved

"*The structure is covered with thin but dense shirting that is glued on. The glued areas are then made waterproof with a collodion coating. The horizontal and vertical control elements are about 1.5 m².*

The arms rest between cushions at 'a-b'. The hands grip at 'ee'. The rear edge of the wings is made of a 1.5-mm cord, with the material glued around it. The willow wood can be obtained from the suppliers of basket makers. I always make the large rod "f" for the steering out of bamboo, which bends slightly when heated above a flame. The other rods can also be made from bamboo, which you may find even easier, as handling the willow wood requires some practice. But bamboo does not have the same toughness as willow wood.

For the rest, trust your practical instincts, and I recommend simply getting started building afresh. Excessive care is not needed. Your first devices will soon break when you start practicing. At first, you should only perform experiments in moderate winds and at heights of 1.5–2.5 m. Choose a sandy spot so that your feet hit the ground softly. Extend your legs far in front so that you don't topple over forwards."

Fig. 17 Landing in the bottom of the pit. *Photo* Kassner, August 7, 1892. © Otto-Lilienthal-Museum.

> *"You could perhaps also build a second device with an area of 15 m², as no experiments can be performed until spring. You need a way to store your wings near the practice grounds after assembling them.*
>
>   *I would be grateful if you would occasionally send me news, and I wish you the best of luck from the bottom of my heart. Your most devoted Otto Lilienthal."*

Lilienthal ended his report on the experiments of 1892 with the observation that the area around Berlin had few suitable training grounds for gliding. As an ideal terrain, he suggested a sand hill sloping in all directions, at least 20 m high, which would allow jumps in every direction. This was also the first time that he mentioned the Rhinower Mountains, around 100 km northwest of Berlin. He said:

> *"Those who feel inspired to practice gliding should note that there is a stretch of land between Rathenow and Neustadt on the Dosse river, the so-called Ländchen Rhinow, which offers a large selection of desirable peaks."*

But Rhinow was even further away than Derwitz. The airfield could only be used on weekends. Accordingly, Otto kept searching for another suitable

Fig. 18 Break from flying in the summer of 1892 at the sand pit in Südende. Lilienthal, his wife Agnes, their sons Otto and Fritz, and Lilienthal's assistant Eulitz. The woodshed where the flying machines were stored can be seen at the edge of the picture on the left. Eulitz was no longer actively participating in flight exercises with the new glider. © Otto-Lilienthal-Museum. All Rights Reserved

location, one that was closer. In 1893, he travelled to the Rhinower Mountains, where he made his longest and most spectacular flights. We do not know how often he used this location for flying. There are eight known photographs taken by photographer Alex Krajewsky, presumably all on the same day.

To perform his flight exercises more frequently, Lilienthal found the so-called "Maihöhe" in Steglitz, a hill close to home. He contacted the owner of the property and, after coming to an agreement, arranged for a self-designed tower-like shed to be built on the hill's summit. This would allow him to jump off the roof with just a few steps. But the Maihöhe was not his final flying location. The wind would break on the walls of the shed, meaning that Lilienthal experienced a surprising amount of turbulence after jumping, requiring great physical effort to overcome. The photographs of his jumps clearly reveal the danger that he faced on every take-off (Fig. 19).

Fig. 19 After jumping from the starting platform on the Maihöhe, Lilienthal has to steer aggressively to correct the course of the glider. *Photo* Anschütz, 1893. © Otto-Lilienthal-Museum.

Fig. 20 Flight practice on the Maihöhe with assistants on the ground. *Photo* Anschütz, 1893. © Otto-Lilienthal-Museum.

Captured in Mid-Air

After Lilienthal's most important photographer, Ottomar Anschütz, had perfected his patented shutter that enabled him to the capture short exposure photographs of the storks that had been so popular, the concept of "instantaneous photography" became extremely popular.

Two well-known companies offering photochemical products like paper and negative film plates in Berlin were "*Agfa*" (*Aktiengesellschaft für Anilinfarben*) and "NPG" (*Neue Photographische Gesellschaft*). Other companies specialized in building cameras and lenses. Several different versions of a quick-release shutter mechanism for instantaneous photographs were presented at a large trade exhibition in the summer of 1896.

But, in order to take these photos, capturing motion at short shutter times, photographers needed a medium that was sufficiently sensitive. At the time, negatives were produced at first on glass plates within the camera, which then could be transferred onto photographic paper in sunlight to create positive images. Initially, the negative process used plates that were treated with collodion, a solution that combined nitrocellulose with ether and alcohol—coincidentally, the same substance Lilienthal used to treat the fabric used on his gliders. These plates needed to be developed almost immediately after the photo was taken, which mean that photographers had to travel with a

Supplementary Information The online version contains supplementary material available at (https://doi.org/10.1007/978-3-030-95033-0_9). The videos can be accessed individually by clicking the DOI link in the accompanying figure caption or by scanning this link with the SN More Media App.

M. Raffel and B. Lukasch, *The Flying Man*, Springer Biographies, https://doi.org/10.1007/978-3-030-95033-0_9

portable darkroom, usually a tent set up at the location of the shoot. By the end of the nineteenth century, dry plates, which were much easier to work with and, more importantly, could be stored and developed much later, became commercially available, though they were far more expensive.

Because the exposed plate needed to be changed after every photo taken, it was impossible for photographers like Krajewsky or Anschütz to take multiple photos of one of Lilienthal's flights. The process to change plates took too long, and the flights were too short. Anschütz had developed a method to capture a series of images in rapid succession using multiple cameras triggered by a mechanical clock. In 1887, he presented his *Elektrischer Schnellseher* (*Electrical Fast Viewer*) which combined pictures taken from 24 cameras into a series that could be viewed by turning a crank. Given the fact that he avidly took images of Lilienthal at each stage of flight, from his takeoff all the way through the landing, Anschütz undoubtedly would have wanted to take moving pictures of the flying man, but, even though he'd used his new multi-camera method to do that with other subjects, it would have been

Fig. 1 Lilienthal in a starting position, presumably by Ottomar Anschütz, Lichter-felde, August 16, 1894. Without support from air currents, the device is very tail-heavy, and it is difficult to hold it straight with the forearms. Lilienthal therefore allows the rear edge of the frame ring to rest on him. In the air, the lift of the wings acts behind the pilot. Lilienthal had to determine the correct position for the pilot in order to trim the glider. When properly balanced, the device should maintain constant attitude with Lilienthal hanging upright within it, with the legs slightly in front of him. (▶ https://doi.org/10.1007/000-6yj) © Otto-Lilienthal-Museum. All Rights Reserved

Fig. 2 Lilienthal at the start of a flight. *Photo* Anschütz, Lichterfelde, August 16, 1894. © Otto-Lilienthal-Museum. All Rights Reserved

Fig. 3 Lilienthal in flight. *Photo* Anschütz, Lichterfelde, August 16, 1894. © Otto-Lilienthal-Museum. All Rights Reserved

Fig. 4 Lilienthal before landing. *Photo* Anschütz, Lichterfelde, August 16, 1894. ©
Otto-Lilienthal-Museum.

Fig. 5 Landed! *Photo* Anschütz, Lichterfelde, August 16, 1894. © Otto-Lilienthal-
Museum.

Fig. 6 Otto Lilienthal's estate also contains the series of images *Flying Crane* for a simple image viewer called a *zoetrope* or "wheel of life" that allowed moving images to be created like a "flicker book". (▶ https://doi.org/10.1007/000-6yh) © Otto-Lilienthal-Museum. All Rights Reserved

Fig. 7 In the *Praxinoscope*, the shutter worked with mirrors instead of viewing slits. © Otto-Lilienthal-Museum. All Rights Reserved

Fig. 8 *Goerz-Anschütz Instantaneous Camera*. © Otto-Lilienthal-Museum. All Rights Reserved

impossibly complicated to configure the timing and orientation of multiple cameras along the path of one Lilienthal's flights.

Anschütz's Schnellseher was a forerunner of cinematography. A number of machines were set up in public places in Berlin at the time and could be set in motion when an interested viewer inserted a coin. The series of images were mounted on a disc which was then turned and becoming moving images. The *Schnellsehers* were produced by *Siemens & Halske* in Berlin and were a special attraction at the 1893 World's Fair in Chicago.

Although cinematography was beginning to emerge around this time, with Max Skladanowsky presenting the first motion pictures to a paid audience at the Berlin Wintergarten on November 1, 1895, no films were ever made of Lilienthal's flight experiments. There is a short silent film showing Lilienthal with his flying machine that could potentially be mistaken for an original recording. However, it was filmed in the 1920s, with aviation pioneer Hans Richter playing the role of Lilienthal.

The company C. P. Goerz in Friedenau near Berlin offered an improved form of Anschütz's patented instantaneous camera made from polished walnut. The exposure time was configured with a cord that ran around the slit when the camera was opened (Fig. 8).

This device was further developed into a folding camera at the end of the nineteenth century, with Ottomar Anschütz's name, city, and province

Fig. 9 Improved folding camera, which became the standard for cameras for many years. © Otto-Lilienthal-Museum. All Rights Reserved

printed on the back of the rolling shutter. The exposure time could now be adjusted from the outside. It weighed 3.3 kg, including six plates and a camera bag. The image could be cropped and framed with a fold-out visor beforehand. A contemporary report claimed that using Anschütz's instantaneous camera was so easy that anybody could do it without the slightest prior knowledge (Fig. 9).

In addition to Anschütz and Kassner, Lilienthal's other photographers are also known by name.

Fig. 10 Flight on the Maihöhe. *Photo* Anschütz. © Otto-Lilienthal-Museum. All Rights Reserved

Fig. 11 Flight in the Rhinower Mountains. *Photo* Krajewsky. © Otto-Lilienthal-Museum. All Rights Reserved

Alex Krajewsky, the photographer who captured the spectacular images of Lilienthal's experiments in the Rhinower Mountains in the summer of 1893, was the "court photographer" for Prince Aribert of Anhalt, a duchy of the German Empire. Later, he opened a studio in the northwest of Berlin and worked as a military photographer for a high-ranking commander during WWI (Fig. 11).

Lilienthal's estate contains one especially interesting photograph that was never published. It shows Lilienthal in free flight high above Spandau, the photographer's hometown. It's a striking image, but also impossible, given the altitude and location. The photo is a montage that Krajewsky had created by taking one of his photos of Lilienthal from the Rhinower Mountains and superimposing it onto a landscape photo of Spandau. Given the challenges of working with wet plates in photography at the time, a montage like this would have been extremely difficult to create. Krajewsky presented it to Lilienthal, possibly as a joke, perhaps including it with an order of other photographs, or he may simply have wanted to show off his skills (Fig. 12).

Krajewsky also photographed Lilienthal at the *Fliegeberg* in Lichterfelde from 1893 to 1895. Some of photographs from Krajewsky's studio are signed

Fig. 12 A spectacular photographic montage: Lilienthal in flight high above Spandau. © Otto-Lilienthal-Museum. All Rights Reserved

Fig. 13 Another flight in the Rhinower Mountains, Alex Krajewsky, signed Henry Magel, Rhinow 1894. © Otto-Lilienthal-Museum. All Rights Reserved

by the photographer Henry Magel, who worked for Krajewsky from 1892 to 1895. These are presumably later prints, and are of better quality than the originals. Magel, who meanwhile had opened his own studio in Ilsdorf near Mücke in Hessen, appears to have kept the original negatives in his possession and was able to reprint them later in better quality on what was called "develop out paper" (DOP), a photographic process that used chemicals rather than light to develop an image in his darkroom.

Another photographer of Lilienthal was Dr. Richard Neuhauss from Berlin, a general practitioner who was also an anthropologist and sponsor of scientific photography. Neuhauss photographed Lilienthal on at least four days in 1895. Krajewsky, a Russian photographer named Preobrazhensky, and another photographer named Regis also took photos at the *Fliegeberg* in 1895. Neuhauss was probably present at some of their photography sessions as well, based on some of the images that have survived (Fig. 6 on page 162).

Lilienthal's essay *Flying Sports and Practical Flying*, published in *Prometheus* in 1895, bears the annotation (Fig. 15):

> *"The photographs were taken by Dr. Neuhauss and Dr. Fülleborn with the Stegemann secret camera designed by Dr. Neuhauss."*

Fig. 14 Lilienthal in flight with the small biplane at the *Fliegeberg*. *Photo* Neuhauss, October 7, 1895. © Otto-Lilienthal-Museum. All Rights Reserved

Fig. 15 One of the photographs published in the *Prometheus* magazine. © Otto-Lilienthal-Museum. All Rights Reserved

Little is currently known about Fülleborn, but the Stegemann so-called secret camera is named after a Berlin-based company called *A. Stegemann*. At the time, a "secret camera" was a type of lightweight and outwardly inconspicuous handheld camera that allowed snapshots to be shot quickly and almost unnoticeably. The Stegemann company, originally a cabinet maker, was said to be unrivalled in building wooden cameras.

Stegemann's secret camera was equipped with an unusually bright lens with a focal length of 10.5 cm made by the *Zeiss Company*. The width of the rolling focal-plane shutter could be adjusted between 0.2 and 8 cm. The shutter was remarkably quiet and smooth to operate. In conjunction with the *Zeiss* lenses, it achieved short exposure times even under limited lighting conditions. For example, Neuhauss captured his photographs of one of Lilienthal's gliders on May 29, 1895, between six and half past six in the evening on the *Fliegeberg*. He used 9 × 12 cm plates and usually took his photographs freehand, without a tripod.

Neuhauss was likely another one of Lilienthal's acquaintances from the VFL. In 1892, he presented images of clouds to the association that he had captured from the ground. After 1894, he served as editor of the magazine *Photographische Rundschau* (*Photographic Review*) for almost a decade and a half, as well as the deputy chairman of the *Freie Photographische Vereinigung* (*Open Photography Association*) in Berlin. In 1895, he published two photographs of Lilienthal's flights in the *Photographische Rundschau*. In November of the same year, he presented photographs of Lilienthal's biplane flights to a large audience at a projection evening organized by the *Freie Photographische Vereinigung* in the Museum of Ethnology.

In 1901, Neuhauss moved into Lilienthal's house on Boothstrasse 17 in Berlin-Lichterfelde as a tenant after the latter's death until the construction of his own house on the nearby Marienstrasse was complete. He died in 1915, and his grave in the Lichterfelde garden cemetery has been preserved.

A few private photographs of Lilienthal are also known from the studio of the photographer A. Regis. His studio was on the Prinzenstrasse in Berlin, not far from Lilienthal's factory. His name first crops up in connection with Lilienthal in 1889. At the time, Otto Lilienthal was preparing the third lecture in his cycle on *The Power Expenditure Required for Bird Flight and its Impact on the Feasibility of Free Flight*. The lecture was delivered to the VFL on April 15 of the same year in the hall of the Royal War Academy. On February 25th, Otto's brother Gustav wrote to his fiancée Anna: "*He will present experimental proof that the wind is directed upward with instantaneous photographs. I suggested Regis as a photographer.*"

Fig. 16 The portrait photograph in the style of a carte de visite on the back of the exhibitor's pass for the industrial exhibition 1896. Photograph by Regis. © Otto-Lilienthal-Museum. All Rights Reserved

It is not known whether the planned snapshots ever came to fruition, but it seems unlikely since they are nowhere to be found. Regis is also the one to capture the well-known last portrait of Lilienthal. The widely circulated version of the photograph was retouched, but an unedited print is preserved at the *Otto-Lilienthal-Museum*. The back of the carte de visite was used as a pass for exhibitors at the Berlin industrial exhibition in 1896, where Lilienthal's company displayed steam boilers and steam machines. The portrait was used in an article entitled *The Flying Man* in the United States and Great Britain in 1894 (Fig. 16).

There is another interesting feature of Regis' photographs. The fact that early photographers used what was called "print out paper" (as opposed to the chemical process mentioned earlier) meant that it was only possible to make

Fig. 17 An early enlargement, Regis, October 7, 1895. © Otto-Lilienthal-Museum. All Rights Reserved

Fig. 18 The corresponding original photograph. © Otto-Lilienthal-Museum. All Rights Reserved

direct copies of photographs, not enlargements. However, Regis produced copies of three photographs showing Lilienthal in flight over the *Fliegeberg* in Lichterfelde where the glider fills almost the entire image. The only way to make such an enlargement would be to make a copy of the negative, presumably using something like burning acetylene or magnesium as an artificial light source. This interesting detail demonstrates that Lilienthal's photographers were themselves pioneers in their field. A contemporary article gives a description of a "mechanical" enlargement process wherein the photographic layer is detached from the glass slide with a tincture and applied to a new glass slide in an appropriately enlarged "expansion bath" after treatment (Figs. 17 and 18).

One of the two pictures of Lilienthal's crashed aircraft in the yard of his factory in August 1896 is attributed to Regis, and he presumably took the other one as well. It is an elaborate print, probably a platinum print made as a copy of a mounted gelatin silver print. These photographs were likely officially ordered as part of the police investigation into the accident.

Other valuable photographs were taken by a Moscow mathematician named Pyotr Vasilyevich Preobrazhensky. He published works on chemistry, technical physics, and photography. The content of his photographs from 1895 at the *Fliegeberg* is especially interesting. The images show details of the control mechanisms that Lilienthal was testing on a large experimental monoplane. His photographs only became known in Germany in 1961, when the *Deutsche Flugtechnik* (*German Flight Technology* magazine) reprinted two articles by Prof. Nikolai Yegorovich Zhukovsky, a renowned Russian flight theorist, from 1896 and 1897. These articles were illustrated with many photographs of flight, three of which were attributed to Preobrazhensky by Zhukovsky (Fig. 19).

An American physicist named Robert W. Wood was the only one taking pictures of Lilienthal's flights in 1896. Wood travelled to Berlin in 1894 after studying the natural sciences and became an assistant to Prof. Heinrich Rubens, a well-known physicist at the university. A few years later, he would himself become a professor of experimental physics at John Hopkins University in Baltimore, Maryland. His research into the field of optics made him famous.

On August 2, 1896, Wood succeeded in taking three photos of Lilienthal's flights in 1896 with the large biplane at the Gollenberg near Stölln. As well as being the only photographs of Lilienthal at his new airfield on the Gollenberg, the hill next to the Hauptmannsberg where he had previously flown in 1893, they were the last photographs ever taken of Otto Lilienthal. Just one week later, Lilienthal's fatal accident occurred at that very location.

Fig. 19 Lilienthal's *Experimental Monoplane* from 1895 (in Germany frequently called *Vorflügelapparat*). The resistance surfaces, which act as wing tip rudders (or spoilerons) for controlling the aircraft, are not shown mounted in all other photographs of the glider, but are clearly visible here. © Otto-Lilienthal-Museum. All Rights Reserved

Fig. 20 Otto Lilienthal in flight with the *Large Biplane*. *Photo* Wood, August 2, 1896

Like so many others, Wood probably also made Lilienthal's acquaintance through the VFL. According to his own statements, he had been interested in the question of flight for a long time. He visited Lilienthal in the factory on August 1, 1896, then drove with him to Stölln the following day.

Modelled After a Bat—The Path to Aircraft Production

In 1893, the third year of his practical flight experiments, Otto Lilienthal abandoned both the large wingspan and rigid design of his previous gliders. His new designs were foldable which made them easier to transport and store. Most importantly, however, the curvature of the wing could be changed as needed using insertable rods, and later sliding profile rails. These benefits were enough to prompt Lilienthal to abandon the sparred wings, even though they were better from an aerodynamic standpoint. The glider could now always be handled by one person during transport, and it was just two meters wide when the wings were attached, meaning that it could fit through doors without any problem.

Lilienthal was not yet sure about the best curvature for the wings, and he was experimenting with different profiles. He was now able to adjust the shape of the airfoil before every flight by inserting a different set of profile rails. With the 14 square meter gilder, he achieved the best lift when the curvature height was 1/18 to 1/20 of the chord. This was surprising to him, as in his book on bird flight he had recommended a ratio of 1:12 based on his measurements from his rotary airfoil testing device.

His oral annual report was once again published in the *Zeitschrift für Luftschifffahrt*, the journal of the VFL. His articles were also used and translated by Octave Chanute for his article series *Aeronautics*, as well as for the French magazine *L'Aeronaute*. This year Lilienthal also published an article about his flights in the popular science magazine *Prometheus* for the very first

M. Raffel and B. Lukasch, *The Flying Man*, Springer Biographies, https://doi.org/10.1007/978-3-030-95033-0_10

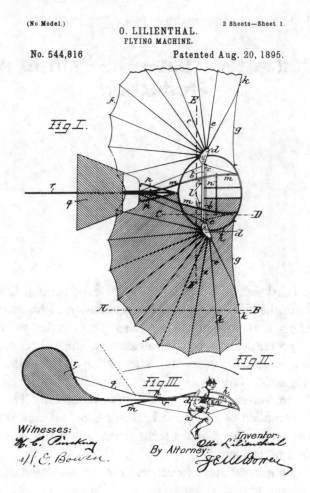

Fig. 1 The basic structure of the collapsible glider in the U.S. patent application from 1894

time. This meant that he reached a much larger audience than the association's journal. Both articles were illustrated with photos by Anschütz. His report described his new flying machines as follows:

> "*As a special innovation for my devices this year, I introduced a way to collapse them. The wings are made from radially arranged ribs and can be folded like a bat's wing. This results in better transportability and allows them to be stored in any space.*"

According to Lilienthal, the new wingspan was no greater than 7 m, and the chord was at most 2.5 m. This gives a wing surface of 14 square meters and a weight of 20 kg. With these dimensions, Lilienthal was once again

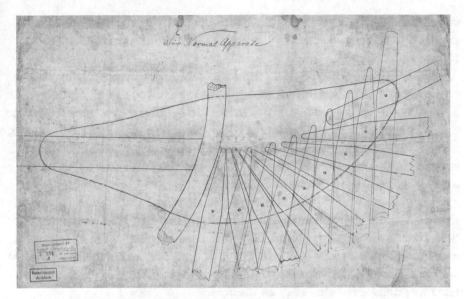

Fig. 2 The heart of the collapsible design: the hinges could only be produced from detailed templates at a scale of 1:1 (preserved workshop drawing). © Otto-Lilienthal-Museum. All Rights Reserved

Fig. 3 The device was easy to transport when folded up. Lilienthal kept the same design for all later constructions that he flew. The photograph published in *Prometheus* is the only one that shows his ornithopter with the engine installed. *Photo* Anschütz, August 16, 1894

able to safely control the flying device in the air even under irregular winds, simply by adjusting his center of gravity to quickly restore balance whenever something happened.

Earlier, in the spring of that year, Lilienthal had set up a "flying station" on the Maihöhe in Steglitz, around 500 m west of his test site in Südende. The flying station consisted of a tower-like shed at the edge of another incline that had been left behind by sand excavation. The shed served as a hangar, allowing Lilienthal to store his machines, and as a Launchpad—Lilienthal would jump off the sloping roof. Together, the shed wall and slope formed a jump height of 10 m, just as in Südende.

The embankment on the Maihöhe also sloped toward the west, and the shed's grass roof sloped in the same direction. However, if Lilienthal thought that he had managed to account for the most frequent wind direction, he was wrong this year too. In the first half of the summer, the wind alternated almost exclusively between northerly and easterly directions, and the tower remained unused for almost three months.

The best flights achieved on the Maihöhe went over a distance of about 50 m. But every time that Lilienthal jumped off the shed roof, he was hit by a violent gust of wind from below, threatening a catastrophic accident more than once. Some of Anschütz's photographs by show Lilienthal in rather critical flight status.

Even though the design changed over time, the materials remained the same. The frame of the glider was once again made from willow rods and covered with shirting. Each wing was formed by seven ribs that radiated out from an articulated pocket on the inner circle of the frame. The ribs spanned the wing radially like an umbrella from the hinge pockets on the frame ring. This design was gradually improved over time. The first rib was stabilized with an outrigger. Later, the number of ribs was increased to nine, and the wing shape was fixed with the much more stable T-shaped profile rails.

Each pocket was fitted with a "spreader", a small vertical brace attached to the upper tension cords. The upper bracing was not exposed to any force during flight, it only served to fix the shape of the wing on the ground and during assembly. The lower bracing, on the other hand, was made of galvanized wire and was subject to load during flight. At the bottom, the tension wires connected to the outermost point of the cross frame. The vertical tail surface had a rounded shape, as did the horizontal stabilizer in front of it, a change from the previous triangular design.

Anyone who admires Otto's bravery during these early flight attempts must surely be even more impressed by the courage of his wife Agnes, who can

Fig. 4 Otto Lilienthal in flight at the *Fliegeberg*. The unusual position of the photographer makes the profile rails easy to see. When the photograph was taken in 1895, the horizontal and vertical fins have already been combined into a cross-shaped tail unit. *Photo* Neuhauss, 29.06.1895. © Otto-Lilienthal-Museum. All Rights Reserved

Fig. 5 The illustration, published by both Pilcher and Zhukovsky, shows the glider that was sold to the Englishman Bennett in 1895 and can now be found in the *Science Museum* in London. The lower bracing attached to the cross frame is clearly visible

Fig. 6 The sloping roof of Lilienthal's "flying tower" on the Steglitzer Maihöhe was protected with a railing at the back. On the roof to the left, Lilienthal's current assistant, Rauh. On the right, Lilienthal's 14-year-old son Otto. To distinguish him from his father, the family called him *"Otto Two"*. Lilienthal stored his gliders in the shed. *Photo* Anschütz, 1893. © Otto-Lilienthal-Museum. All Rights Reserved

be seen in many of the photographs. In the first few years, with an estimated number of several thousand flights, very few were entirely without moments of danger. Agnes was no doubt fully aware of this, as she often attended and watched the experiments. Despite the concern she must have felt for her husband, and although she never expressed any ambition nor satisfaction regarding his growing fame, she always believed that Otto would be successful. The Lilienthal children also often visited the Maihöhe while their father was flying. It was exciting to watch him slowly but surely achieve distances of up to 50 m. The children can also be seen in many of the photographs from this period.

Fig. 7 Anschütz documented every phase of flight. © Otto-Lilienthal-Museum. All Rights Reserved

Fig. 8 After jumping, Lilienthal often needed to correct for turbulence in the wind at the shed's wall. Otto-Lilienthal-Museum. All Rights Reserved

Fig. 9 In stable flight position. © Otto-Lilienthal-Museum. All Rights Reserved

Fig. 10 Lilienthal in flight in front of a panorama of the village of Steglitz. This picture was used in 1895 as an advertisement for the gliders manufactured by Lilienthal's factory, making it the first ever sales advertisement in the history of aviation. © Otto-Lilienthal-Museum. All Rights Reserved

Segelapparate

zur Uebung des Kunstfluges

fertigt die Maschinenfabrik von

O. Lilienthal

Berlin S. Köpenickerstrasse 113.

Fig. 11 The first advertisement for an aircraft in history

Fig. 12 Contemporary print of a photograph of Lilienthal's flights by Ottomar Anschütz. As was customary at the time, the prints were mounted on strong cardboard, which also served as a business card for the photographer. Anschütz's photograph cards were decorated with a gold rim and always carried the inscription: "*Reproduced true to life by Ottomar Anschütz, Lissa (Posen)*". © Otto-Lilienthal-Museum. All Rights Reserved

Fig. 13 Again, a strong correction of the flight by shifting the body weight. The glider tilts toward the ground to the left. Lilienthal has to react counterintuitively by throwing his legs and body to the right to press against the raised right wing. The natural reaction here would be to shift one's feet to the left to absorb the expected impact on the left surface. *Photo* Anschütz, 1893. © Otto-Lilienthal-Museum. All Rights Reserved

Fig. 14 During the approach for landing: again, the pilot must react counterintuitively. Lilienthal is still gliding. However, if his feet come any closer to the ground, he must throw them backward to lift the nose of the glider. This slows down the flight, breaking the air flow and allowing him to touch down gently. After hitting the ground with his feet, he comes to a stop after a few quick steps. If he raises the front of the glider, the horizontal tail surface flaps passively upward and does not obstruct the straightening process. Lilienthal liked to demonstrate how expertly he had mastered the landing process by landing on one leg and immediately coming to a stop, according to eyewitnesses. *Photo* Anschütz, 1893. © Otto-Lilienthal-Museum. All Rights Reserved

Practical Flying—Training and Records

In 1893, Otto discovered another promising new location for his flights. The hills near Rhinow in the Mark Brandenburg were up to 60 m high, with very little vegetation at the time.

> "*I have now moved my primary practice area to the Rhinower Mountains, which I mentioned in my report from last year. The terrain here seems almost intentionally made for flight experiments. A chain of hills up to 60 meters high rises above the surrounding flat fields, covered with grass and heather, offering inclined slopes in every direction.*
>
> *The slope of the hillside varies between 10 and 20 degrees, and there is plenty of choice for which way to glide down them in the air. When I unfolded my flying gear on these mountain slopes for the first time, I was admittedly overcome by a somewhat anxious sensation when I said to myself: 'from up here, you must now glide out into the vast expanses of land lying far below!' But after just a few careful jumps, I soon regained my sense of safety: gliding was much gentler here than from my flying tower. The wind did not rise up like it did at the tower, where every time I crossed the edge to jump I would be battered by an uneven gust of wind from below, often threatening to provoke a catastrophic accident.*"

These are the words chosen by Lilienthal in his annual report of 1893 to introduce his flight tests in the Rhinower Mountains, a chain of hills extending from east to west some 15 km north of the town of Rathenow through the flat Rhin-Havel-Luch region. Remarkably, this is the very first

M. Raffel and B. Lukasch, *The Flying Man*, Springer Biographies, https://doi.org/10.1007/978-3-030-95033-0_11

Fig. 1 Another flight in the Rhinower Mountains. This photograph was printed multiple times, including in the article "The Flying Man" in the English newspaper *Westminster Budget* in 1894 and the American periodical *McClure's Magazine*. In *Prometheus*, Lilienthal wrote: *"This photograph was taken when I was being lifted by the wind to a greater height and was stalled in the air, as if rooted, because the sudden strength of the air current carried me and prevented me from advancing forwards."* Photo Krajewsky, 1893. © Otto-Lilienthal-Museum. All Rights Reserved

time that Lilienthal or anyone used the word *Flugzeug* to describe his glider. Analogous to the term *Fahrzeug* (vehicle, but more literally "driving gear"), Flugzeug directly translates in English to "airplane". This is a revealing choice of name for the collapsible, fabric-covered wings. In their builder's eyes, Lilienthal's aircraft were designed for recreation, like any other sports or fishing gear. Nobody suspected that the same word would be used in the distant future to refer to machines that transport hundreds of people around the world simultaneously. At the time, the English word "flyer" and the French words "avion" and "aéroplane" were well-established. In Germany, early models of powered aircraft were initially called *Drachenflieger* (hang gliders). The term *Flugzeug* only became widely used for flying machines in 1910, long after Lilienthal's death, when a special commission was established to clarify aeronautical terminology. The commission endorsed *Flugzeug* while rejecting many other terms that had been proposed or were already widely used.

Fig. 2 Lilienthal wrote the following about this photograph in *Prometheus*: "*I was coming from a hill to the right whose foot is visible in the picture and flew toward the plain by making several turns. The photograph was taken at the exact moment when I almost turned my back on the plain.*" © Otto-Lilienthal-Museum. All Rights Reserved

Over the summer of 1893, in the Rhinower Mountains, Lilienthal continued to use the same glider that he had transported by train from the Maihöhe to Rhinow. The Rhinow Mountains are located between the railroad stops at Rathenow in the south and Neustadt on the Dosse river in the north, and they became Lilienthal's favorite practice site in the summer of 1893. In the beginning he mainly flew from the Hauptmannsberg peak at a height of 95 m, then around Easter 1894 and again in 1896, he practiced on the Gollenberg near Stölln to the east.

In 1893 Lilienthal frequently stayed at the *Zum Rhinower Ländchen* inn, and this is probably where the glider was also stored. Beginning the following year, when he started flying on the higher Gollenberg, he regularly stayed at the *Herms Inn* in Stölln, a few kilometers east of Rhinow.

Lilienthal soon reached distances of up to 250 m, staying in the air for about 30 s. These gliding flights, flown along a slope, were still part of the curriculum for German glider pilots until the 1950s. Lilienthal also became more proficient at controlling the glider, successfully performing a

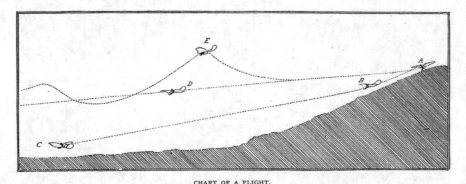

CHART OF A FLIGHT.

A. The start. *B.* The gliding descent. *C.* Alighting in still air. *D.* Course in ten-mile breeze. *E.* Soaring in a strong breeze.

Fig. 3 A drawing by Lilienthal of his flight lines in the Rhinower Mountains in *McClure's Magazine*: Chart of Flights A: *The start*. B: *The gliding descent*. C: *Alighting in still air*. D: *Course in ten-mile breeze*. E: *Soaring in a strong breeze*. © Otto-Lilienthal-Museum. All Rights Reserved

full 180-degree turn. It must have been an incredible feeling to see his flight performance improve so greatly. While presenting to the VFL, he said:

> "*It does not take long until the flyer ceases to care whether he is 2 meters or 20 meters above the ground; he can feel how safely the air is carrying him, even when there are tiny people on the ground looking up at him in amazement. Soon, he is crossing gorges as high as a house and sweeping hundreds of meters through the air without any danger, successfully parrying the wind at every moment.*"

When experimenting with turns, Lilienthal started small, making gentle deviations to the left and right by shifting his weight side to side. This change in the center of gravity causes the wing to enter a bank, initiating a turn in the desired direction. "*There is nothing simpler than piloting a flying machine*," he once exuberantly exclaimed.

The fasted wind speeds in which Lilienthal dared to fly in Rhinow were 7 to 8 m/second. With winds this strong, he and his glider would sometimes come to a standstill in mid-air, hovering over the same point for seconds at a time. Occasionally, he reported that, while hovering, a sudden gust would send him straight up several meters, which could be quite frightening. But he always managed to regain control and land safely. He had the following advice for new would-be pilots:

> "*Be especially wary of the following mistake: suppose the glider is hovering in the air and suddenly feels himself being lifted by the wind, typically unevenly, for example*

with the left wing lifted higher than the right. This slanted position will drive him toward the right. A newcomer might involuntarily extend his legs to the right, anticipating an impact on the ground there. But this increases the load on the already lowered right wing even further, causing the flight to quickly plummet to the right, until the right wing buries itself in the ground and buckles. There is little danger to life or limb; the glider forms an effective collision frame that absorbs the force of the impact. The correct response is to extend the legs toward the raised wing at all times to push it back down. It of course requires some practice, but this useful movement soon becomes automatic once you have convinced yourself how safely it allows you to steer the wings and protect them from being destroyed."

For the time being, Lilienthal did not consider it advisable to place his body in a stretched-out position and fully entrust himself to the glider while lying down, as the Wright brothers did a few years later. Lilienthal wanted to keep his legs free to run, jump, steer, and land at all times as a precaution in case he lost control. In the future, he said, one might be able to switch to a lying or sitting position within the glider, once they had become much more sophisticated. He was fully aware that hanging from the glider increases the air resistance considerably, and that any other position would be greatly beneficial from an aerodynamic standpoint. But he did not yet see a way to achieve this safely.

As early as 1893, Lilienthal had thought about using some kind of *wind gauge*, what we'd now call an airspeed indicator, on his glider, as maintaining sufficient speed (relative to the surrounding air, not the ground) is a key factor in flying. While the only "airspeed indicator" he had was the feel of the wind on his face, he thought about some kind of mechanical solution as well.

"A small display mechanism on the front of the glider that allows one to read the relative motion of the air at all times would be very useful. But this mechanism must not cause any significant increase in air resistance."

At the end of his annual report, Lilienthal explicitly expressed the wish for his publications to encourage those who were studying aeronautics from a theoretical standpoint to actually get out and try it, performing practical experiments themselves. He emphasized that he himself adhered to only one of several principles of flight, namely the imitation of birds. He was convinced that this approach was logically correct and offered great advantages. But he did not object to other concepts, such as helicopter or propeller-based flight, as such. Only practical experimentation can ultimately reveal which concept will lead to success, he said.

Lilienthal spent many of his weekends in the summer of 1893 at Rhinow. He would typically catch an early suburban train from Lichterfelde to Berlin at 4 a.m. on a Saturday, followed by a long-distance train to Hamburg at Lehrter station, arriving at Neustadt on the Dosse in a little less than an hour and a half. Then he would ride a horse-drawn cab south to Rhinow. This final stretch was around 15 km, meaning that the entire journey took about four to five hours. Lilienthal was usually accompanied by his technician Rauh, who had replaced Hugo Eulitz to help with his experiments. But in May, Lilienthal hired Paul Beylich in his factory, a young locksmith who would serve as his chief engineer and assistant. From this point on, Beylich built all of Lilienthal's flying machines. Beylich later said that Lilienthal did not fly once after 1894 without Beylich being present. He was also the only eyewitness to Lilienthal's fatal crash in Stölln in August 1896.

The location in Rhinow was ideal for flying. But the fact that it was only available on weekends was a major drawback. During the week, Lilienthal was needed in his factory and was unable to travel to Rhinow. He needed another airfield closer to his home in Lichterfelde, so that he could fly much more frequently.

"If these hills were in the immediate vicinity of Berlin, flying would undoubtedly develop into a regular sport; no other sporting trend can compare with the wonderful and effortless sensation of gliding through the air."

Perhaps this would be enough to answer all of the unsolved theoretical questions once and for all:

"It does not seem impossible that simply continuing exercises such as this might ultimately lead to free and continuous soaring flight in windy conditions."

At the time when Otto was describing his "flying gear" as sports equipment, cycling was another area where modern sports and technology were intertwined. In the late nineteenth century, the invention of radial-style ball bearings led to the development of what we would recognize today as modern bicycles, replacing the impractical high-wheel penny-farthings of the 1870s. Lilienthal frequently compared his flying machines to bicycles, citing the latter as an excellent example of the relationship between theory and practice.

"The problem of travelling on two wheels, which our ancestors would surely have scoffed at in disbelief, now appears to have been completely solved. Now imagine that a theory of balancing on two wheels had been developed by somebody before anybody ever thought to implement this skill in practice. Such a theory does of course

exist; it specifies how any cyclist must act in order to maintain his balance. It simply requires the cyclist to continuously turn his bicycle toward the side where he wishes to fall. Thus, in a sense, he always keeps the center of gravity of the bicycle under him; in the case of curved motion, the direction of the force obtained by averaging gravity and the centrifugal force takes the place of the center of gravity. This theory is relatively simple, and it is well-known that riding two-wheeled bicycles is much easier than it seems at first glance, because this theory translates so well to flesh and blood, and following it becomes entirely mechanical after a few days of practice.

But imagine that somebody had proposed a correct theory clearly and distinctly before the two-wheeled bicycle had ever been developed in practice. Would we have realized that the two-wheeled problem was solved? Never! It is essential for practical experimentation to provide the decisive impulse."

Lilienthal also compared the appeal of cycling and its associated costs to his flying machines. An interesting anecdote in the margins of the history of aviation is that Octave Chanute was also thinking about which skills would be necessary to continue the development of aviation, concluding that bicycle mechanics would be well equipped to become capable aircraft

Fig. 4 The Lilienthal family's residence at Boothstrasse 17 in Lichterfelde, September 1887. Otto Lilienthal's son is on the left with a penny-farthing bicycle, and his other children Anna and Fritz are on the right. © Deutsches Museum.

Fig. 5 Garden view of the house, 1887. © Freital Municipal Collection. All Rights Reserved

designers. Chanute's prediction, of course, turned prophetic with the success of the Wright brothers.

Otto's children owned penny-farthings themselves, as can be seen in a photograph of their house. Otto wasted no time in acquiring the new bicycles for his family and rounded all of the corners of their yard to make it easier to ride at home. Naturally, the children would also cycle through their quiet neighborhood streets, though not all their neighbors approved of this modern hobby, especially for girls.

The Lilienthal families were open-minded, with wide ranging interests. Gustav returned from Australia as an unconventional free thinker, wearing a wardrobe consisting of what he called "reform clothing." His wife and five daughters grew up with a relaxed philosophy, eschewing convention and traditional standards of dress, while Gustav himself invented technologically inspired toys that his daughters enjoyed. Otto's family was no different, and his neighbors described him as genuine and down-to-earth, without a hint of arrogance.

"*There were no days off for him,*" said one of Gustav's sons-in-law about Otto. "*His happiness radiated from his daily life.*" Besides the ongoing flying, there were often surprises around the Lilienthal household, like one his impromptu French horn solos at the window.

One day, Otto was summoned to his daughters' high school to talk to the headmistress. Neither of his girls would be moving on to the next grade level, she explained, to their father's astonishment. Since the girls had always done well at school, Otto was at a loss to explain the change in performance. The reason, continued the headmistress, was that Anna and Helene had been spending their time cycling through every inch of Lichterfelde on their new bicycles. Not only was this responsible for their allegedly poor academic performance, it was, according to the headmistress, extremely inappropriate behavior for two young girls. When Otto returned home after meeting with the headmistress, Agnes naturally asked about what had happened. After he told the story, Agnes was as stunned as he had been to learn that little Anna and Helene were going to be held back and forced to repeat a grade. "*So, what did you say to the headmistress?*" Agnes asked. Otto responded with a beaming smile: "*Let them keep cycling.*"

Despite their "*heinous*" cycling problem, both girls successfully completed high school. Anna went on to attend foreign language seminars at the University of Greifswald, again achieving good results. All of the children were encouraged to participate in sports. By then, horizontal and parallel bars had been installed in the yard, as well as a small running track. In summer, they swam; in winter, they skated on the ice. Anna practiced diving into the water from the 10-m tower, as well as somersaults from slightly lower springboards, following in the footsteps of her grandfather, who had also had a passion for acrobatics.

Otto's unwavering belief that man would master flying and that it would become a sport comparable to cycling and swimming was frustrated by the fact that he had few peers with whom he could discuss the topic objectively. He knew that the idea of human flight would be perfected if enough people saw the benefits, and he was convinced that recreation would be the catalyst.

"*The key is therefore to find a method that allows flight attempts to be safely organized and simultaneously exploited as an attractive form of entertainment for sports enthusiasts. [...] The appeal of these flights is indescribable. It is impossible to imagine a healthier form of outdoor exercise or a more stimulating sport. Competition in these exercises would undoubtedly lead to the equipment being continuously perfected, just as we saw happen with bicycles, for example.*"

Even when he had just begun making his first flights of just 25 m near Derwitz, he confidently wrote:

"It is more than likely that later generations will be amused by the clumsiness with which man's ability to fly evolved. One day, the simplest of machines will perhaps allow mankind to fly."

Two years later, he vented his disappointment with the rate of developments in Germany:

"It seems to only be the less well-off who (...) invest their scarce time and even scarcer resources into flight tests, especially in our dear old country of Germany. In other countries, things are slightly better."

He was undoubtedly thinking of France, where the government had provided funding for aeronautical experiments, including supporting one researcher named Hureau de Villeneuve. Villeneuve's attempts had been unsuccessful, among other reasons he was using flat wing models instead of curved airfoils. Otto attempted to convince him that this was a mistake, and even sent him a copy of his book, to no avail.

Of course, there were some flight enthusiasts with plenty of their own resources to spare. One such individual was the prolific inventor Hiram S. Maxim, who was able to afford designs with a wing area of 540 square meters powered by a pair of 360-horsepower steam engines. But money wasn't everything, and by no means guaranteed success. Maxim's very first attempted takeoff ended badly. Otto's advice to approach the development gradually and his warnings against attempting to build everything at once had fallen on deaf ears. In 1893, he wrote the following about such designs:

"Progress toward accomplishing the fastest of all methods of transport is proceeding alarmingly similar to the gait of the snail, the slowest of all animals!"

Nevertheless, he believed that *"a very interesting time is soon approaching,"* his natural optimism providing the basis for his utopian vision of aviation's future. When two of Otto's grandchildren, Anna's sons, asked the press to recognize the 50th anniversary of their grandfather's death and the 100th anniversary of his birth in 1946 and 1948, they were met with the response: *"Out of the question. Lilienthal is responsible for terror bombing."* For political reasons the press refused to recognize Lilienthal for his contributions to flight, instead blaming him for the horrific destruction of WWII. Otto himself could never have foreseen such accusations.

An Engine—Again Inspired by Birds

Otto had designed a new aircraft that would pave the way towards powered flight, his first ornithopter, later described as the "small" one. Its design followed a similar principle than the Maihöhe-Rhinow apparatus, featuring folding bat-like wings. In the summer of 1893, Lilienthal decided that the design of his collapsible glider was already sufficiently advanced to apply for a patent. In keeping with his focus on recreation, the patent was filed in the Class 77—Sports category. Lilienthal never imagined the large-scale military and transportation applications of aviation, and those things simply weren't among his objectives. Together with Hermann Moedebeck, the editor of the *Taschenbuch zum praktischen Gebrauch für Flugtechniker und Luftschiffer* (*Pocketbook of Practical Use for Flight Technicians and Aeronauts),* Lilienthal decided to call what he did "personal artificial flight".

The patent, No. 77916, was granted on November 10, 1894 by the *Imperial Patent Office* in Berlin. It summarized the current status of his flight engineering work. The patent drawings show that it was submitted before he decided to merge the stabilizers in a crosswise assembly at the tail. It begins with the words:

"The present flying apparatus is intended to enable free flight by humans, both in the form of gliding flight without wing beats and flapping flight with moving wings. The apparatus consists of a slightly curved surface made from a wooden frame with a fabric covering."

© The Author(s), under exclusive license to Springer Nature
Switzerland AG 2022
M. Raffel and B. Lukasch, *The Flying Man*, Springer Biographies,
https://doi.org/10.1007/978-3-030-95033-0_12

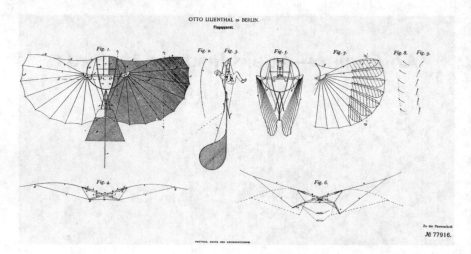

Fig. 1 Lilienthal's collapsible glider in his patent drawing of September 1893. In the patent drawing, each wing is formed of nine ribs (longerons). The passively flapping moving tail surface is also shown in the drawing in Fig. 1 on page 114

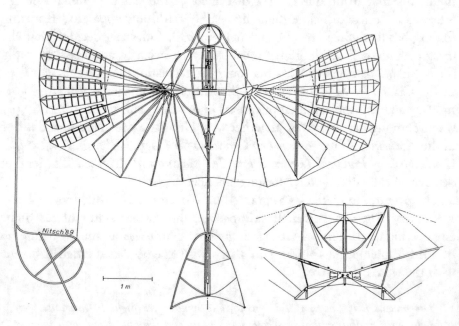

Fig. 2 Reconstruction drawing of the *Small Wing Flapping Apparatus*, Nitsch. © Otto-Lilienthal-Museum. All Rights Reserved

Fig. 3 Otto Lilienthal with his *Small Wing Flapping Apparatus*. *Photo* Anschütz, August 16, 1894. © Otto-Lilienthal-Museum. All Rights Reserved

Specifically, Lilienthal's patent claims covered the wing bracing and the collapsible design of the wings, as well as the fixed vertical and upward flapping horizontal tail fin. He also patented a certain mode of operation by machine or human power.

The patent includes nine schematic drawings, four for the normal glider and five for a powered machine. The patent drawings show a design with nine ribs instead of the initial seven. The freedom of the horizontal tail surface to move upward is also portrayed in Fig. 1 on page 114 of the patent drawings with a dashed line. The patent appears to have been written in parallel with the design of the first ornithopter, since some details, such as the position of the engine, match up precisely with how the apparatus itself was implemented.

Lilienthal was unable to obtain a patent for his most important discovery, the curved wing, which dated back to 1874, as an analogous patent had already been granted to the Englishman Horatio Frederick Philips in 1884. Accordingly, the curved surface is mentioned in the text of the patent but is not listed as one of its claims.

Lilienthal obtained an identical patent in Great Britain in 1894, as well as a restricted version in the U.S. in 1895. The U.S. patent, No. 544,816, dated August 20, 1895, only covers the glider, but not the wing flapping mechanism.

By the time the patents were published, Lilienthal had already made significant improvements to his flying machines. The tail surface had been redesigned as well as the wing design. The shape of the profile was now fixed very precisely by four T-shaped profile rails. The pads on which his arms had previously rested were now pockets that fully enclosed each forearm, giving Lilienthal even greater reliability in steering the glider.

Today, it is difficult to judge the completeness of our knowledge of Lilienthal's many flying machine designs. What we know comes from those drawings and sketches that have been preserved, Lilienthal's correspondence, and the many surviving photographs. The sheer number of photographs and the regular photo shoots suggest that Lilienthal was very deliberately taking advantage of this new medium to document his yearly progress in flight development, as well as his feats of flight. Designs that never made it past the development stage or which did not achieve the desired results would not have been displayed for photographers, and pictures were only taken on a few days in any given year. Just as it was—and is—with flying, weather was always a factor for his photographers, and it may well have been that arranging the photo shoots was even more challenging than the actual flying itself. Regardless, Lilienthal was well aware of the effect produced by the spectacular images. In one of his lectures, he sought to moderate the enthusiasm of his audience:

> "*In conclusion I want to ask you not to take my achievements for more than they are. In the photographic pictures, you can see me flying high above in the sky. One can get the impression that the problem is already solved. That is not at all the case. I have to admit that it will still take quite a lot of work to turn this simple gliding into a long-distance human flight. The achievements so far are for human flight, nothing more than what the first insecure steps of a child mean to walking of men.*"

But Lilienthal clearly also derived great pleasure from the photographic *mise-en-scène* of his flights. Pictures showing Lilienthal dressed in his aviator's costume, together with an audience and other characters, offered an impressive setting for the photos that remain a remarkable testament to the history of both flight and photography to this day.

Lilienthal's flying machines were experimental devices. He often adjusted them during testing. In some photos, repairs to the covering are visible. If the trim was not right, the only way to correct it was to change the wing surface (Fig. 4).

Fig. 4 Circular cuts are visible at the rear edge of the wing. Lilienthal seems to have needed to shorten the wing to trim the device. Autotype of a graphic *reproduced from original photographs* in the magazine *All Deutschland*. The photograph very clearly shows the circular cuts at the rear edge of the wing, which are also documented in other photos. The original photograph on which it is based is however not known. © Otto-Lilienthal-Museum. All Rights Reserved

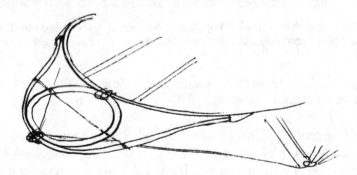

Fig. 5 Sketch of the buffer in Lilienthal's letter to Wolfmüller, December 13, 1894. © Otto-Lilienthal-Museum. All Rights Reserved

Photographs also show that Lilienthal used a clearly identifiable tail unit with different gliders. Another detail is a protective mechanism that he seems to have designed after experiencing dangerous flight situations—a so-called *buffer*, like a crumple zone in modern passenger cars. Lilienthal attached a willow rod bent into a circle to the front of the frame ring as protection. In the operating instructions sent with his glider delivery to Wolfmüller, he describes this element in detail (Fig. 5):

"To protect the apparatus and the body, a buffer can be attached to the front of the apparatus. This buffer should be directed downward at a slight angle to catch

Fig. 6 Lilienthal with the *Small Wing Flapping Apparatus* and a mounted buffer on August 16, 1894. *Photo* Anschütz. © Otto-Lilienthal-Museum. All Rights Reserved

any impact and break before the apparatus itself is damaged. The buffer should be installed as illustrated in the sketch."

There are only a few photographs of Lilienthal that show a mounted buffer. He only developed it to add safety to flights with unproven devices whose flight characteristics were not yet familiar to him, and of course for the buyers of his flying machines. In routine flights with tried-and-tested models, he placed his faith in his skills and his routine, which by now had become very elaborate.

Alongside the ongoing development of both his gliders and his prospective ornithopter designs, Lilienthal, with Beylich's help, was also preparing for series production of his flying machines, something that would certainly be necessary in order for aviation to develop as a sport. On November 14, 1893, Lilienthal wrote a letter to the aviation journalist Hermann Moedebeck:

"The previous publications in which I suggested this notion have aroused the interest of parties in several locations, so you will soon hear more about these gliding exercises. I have created a special factory to manufacture these devices under the charge of a true "aeronautical engineer" whom I have trained specifically for this purpose. The customers who have ordered machines can hardly control their enthusiasm,

Fig. 7 The photographs of Lilienthal are spectacular. The photographer Ottomar Anschütz used one of the photographs he took on this day to advertise his photography studio more than a decade later. © Otto-Lilienthal Museum. All Rights Reserved

Fig. 8 In almost the same position: Lilienthal with his *Normalsegelapparat*. *Photo* Anschütz. © Otto-Lilienthal-Museum. All Rights Reserved

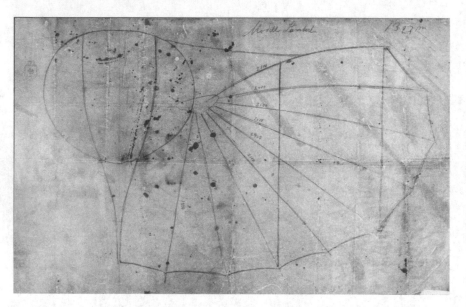

Fig. 9 De Lambert model: drawing preserved in the estate of Paul Beylich. © Otto-Lilienthal-Museum. All Rights Reserved

and some have already sent the requested 300 marks in advance to ensure that they will be served promptly. If everything continues in this manner, the outlook seems quite favorable. It is regrettable that there is not a single hill near Berlin suitable for takeoffs. I must wait until a patron can build one. Next year will undoubtedly bring further progress, as the same flight that is sometimes currently possible at an incline of 1:10 without wing flapping will be flown with powerful wing beats. The machine required for this is largely finished. I believe that I can demonstrate it in a meeting at the factory in the winter, if not in free flight, then at least the flapping wings. You can see in the photographs that I am already quite comfortable in the air. In the hope that I will have the opportunity to demonstrate my experiments to you in person, please accept my respectfully devoted greetings.

Otto Lilienthal"

Some details of the models from the winter of 1893–94 and the summer of 1894 would still be modified, including the size. The numbers written surviving drawings suggest that the specifications still varied from device to device. Perhaps the size and weight of the buyer were incorporated into the design to ensure that the pilot's center of gravity would be correctly positioned relative to the lifting force of the wing (Fig. 9).

Development appears to have been complete by the end of 1894. On a carefully executed final drawing at a scale of 1:10, the flying machine was named the "normal soaring apparatus with a 13 m² gliding area".

Fig. 10 The *Sturmflügelapparat* model in the Technical Museum of Vienna. *Photo* Wittig

From various drawings, we know of other concepts that presumably remained at the design stage, were unsuccessful, or were never realized at all. One such design was a large monoplane model with an area of 17.5 m² intended for flights in calm conditions. Its counterpart was an unusually small flying machine with an area of just 9 m², the *Sturmflügelapparat*. Nothing is known about the large design beyond a draft drawing, whereas the smaller one, whose structure is very similar to Lilienthal's standard glider, was actually built. However, Lilienthal's attempts to use this device to conquer strong winds seem to have been unsuccessful. In any case, there are no photographs of it in its original configuration. Nevertheless, very precise photographic documentation does exist; in 1895, it was converted into a biplane that featured in various photographs. It is interesting that this particular aircraft was ultimately preserved from among Lilienthal's collection. It survived as the *Sturmflügelapparat* alongside the four standard gliders in museum exhibitions and can now be viewed in a restored state at the Technical Museum in the Austrian capital of Vienna (Fig. 10).

In the spring of 1894, Lilienthal tested various airfoil shapes. He experimented with parabolic cross-sections whose curvature was largest near the front edge of the wing, as well as with different depths of curvature. Before Easter, he suffered a flying accident, but escaped without much harm thanks to a bit of luck. Flying what was presumably the "Stölln model", he had recently switched from the Hauptmannsberg to the higher Gollenberg. During one of his first flights, he was forced to stretch his upper body backward to keep the glider from nosing down. He was unable to bend his arms and raise his body from this position, which meant that his weight was extremely far aft and caused the glider to pitch up uncontrollably.

> "*I held on tightly, seeing nothing but the blue sky and white clouds above me, waiting for the moment when the device would tip over backward and perhaps end my gliding experiments once and for all. But suddenly, the apparatus stopped climbing and fell backward from its height, with the diagonally upwards-pointing horizontal tail moving through a short arc until the rear of the apparatus was directed upwards once again, now turned on its head, hurtling vertically down to earth from a height of about 20 meters, myself along with it. With a clear mind, still holding the device by its handles, I tumbled toward the green lawn, arms and head first – a bump, a crash, and I was on the ground with the device. The only dire consequences of my accident were a wound on the left side of the head where I had hit the machine frame and a sprained left wrist. As unbelievable as it may sound, the apparatus was completely intact. My glider and I had been saved by the elastic buffer that I had attached to the front of the device for the first time, as if inspired by some higher providence. The buffer itself, made from willow wood, was completely splintered, and parts of it had buried themselves a foot deep into the ground, requiring great effort to pull them out.*"

To his account of the accident, which was witnessed by his brother Gustav and a group of students and teachers from the Stölln elementary school, Lilienthal immediately added that this was his only fall of this kind out of thousands of glider flights, and that even this could have been avoided with greater caution on his part. Furthermore, such an accident was probably the worst that could occur during a flight, he wrote. Now, one key experience richer, Lilienthal modified the design, moving the hand position further aft and attaching cushions to the frame ring behind him that would support his upper arms in an angled position. This would prevent him from ever again slipping and hanging well below the glider from outstretched arms. At this point, he'd also decided that a profile height of 1:15 to 1:18 relative to the chord would be the best airfoil curvature for the monoplane: safe but strong.

While perfecting his glider, Lilienthal began work on his next goal. Instead of trying to take off from the ground by flapping his wings, as he had done when he was younger, he now wanted to follow the example of the stork. The idea was to try to gain height while gliding by flapping the wings to fly further and stay in the air longer.

When Lilienthal was granted a patent for his flying machine in 1893, the wing flapping device described above was already in development. According to Lilienthal's specifications, the aircraft had a wingspan of eight meters. The number of wing tip fingers, additional fabric-covered surfaces that extended straight out from the wingtips, on each side was reduced from seven to six compared to the specifications in the patent. The maximum amplitude of the wings was 1.2 m at the tips. In the autumn of that year, Lilienthal proposed in an essay that powerful wing beats could be added to his regular gliding. He wrote:

"Once greater dexterity in gliding down from greater heights has been achieved, nothing stands in the way of moving appropriately shaped wings, either with the feet or mechanically, in a such a way as to increase the lift and range of free flight more and more, until permanent horizontal flight is achieved, even if only temporarily under favorable wind conditions."

In parallel, he submitted a program for such exercises to the VFL. Before beginning his experiments with moving wings, however, he needed to become familiar with the flight behavior of his new flying machine. His plan was to first fix the wing tips and only use his new design as a glider until he could operate it as confidently as his previous designs. He expected the wing flapping mechanism to cause the aircraft to behave differently in flight than the gliders, due to the feathered wing surfaces (Fig. 11).

In his patent, he had considered the possibility of wings operated by muscle power. In the figure described in the patent,

"the left side shows the motion of the wings driven by human power by placing the feet in slipper-like leather pockets and causing the motion by pulling in and pushing out the legs. The right side illustrates motion driven by a special machine with two chains running from a piston rod over rollers to the levers. The upstroke of the wing is achieved through suspension of the wing ribs and the air pressure acting from below."

The design of Lilienthal's wing flapping apparatus followed the specifications of the patent very closely. The motion of the wing tips was not induced by joints but by the elasticity of the willow branches. In the lower braces, the wires were not attached directly to the cross frame, but to angled levers that extend it. These levers could be pulled inward with steel cables to trigger a wing beat. The upper braces were elastic, and stretched as the wings beat, pulling the wings back upward, assisted by air pressure from below. The wires could also be attached to the cross frame to create a rigid gliding configuration. Lilienthal described his approach in an article published in October 1894:

"I was therefore forced to content myself with normal gliding flights on this larger and heavier apparatus, which weighed 40 kilograms, twice as much as a regular soaring apparatus, so that I could first practice safe landing with confidence. Only

Fig. 11 The *Small Wing Flapping Apparatus* in gliding flight. The engine has not yet been installed. However, Lilienthal has already installed the buffer to ensure his safety. *Photo* Ottomar Anschütz, August 16, 1894 in Lichterfelde. © Otto-Lilienthal-Museum.

now that this step is successfully complete can I begin to carefully flap the wings in free flight."

Pedal controls for flapping the wings would not have been easy to integrate into Lilienthal's flight technique. He needed his legs for take-off and landing, as well as to control the device during flight. This is no doubt why he immediately began working on an engine.

"We are currently suspending our experiments with rigid wings and have instead begun experiments with moving wings. We started with a new apparatus but were forced to discontinue testing, as the external frame and stability had not yet been sufficiently investigated. This problem has now been resolved, and we are ready to try the wing flapping apparatus once again."

Lilienthal's step-by-step approach had been successful when he was learning to glide, and now he hoped to do the same with powered flight. Although he had experimented with small steam engines on his models in

Fig. 12 Replica of the first carbon dioxide engine by Stephan Nitsch, 1988. © Otto-Lilienthal-Museum.

the past, he chose a different principle to power his flying machine: a single-cylinder compressed carbon dioxide-powered engine that he had designed in 1893. One of the drawings by his technician Hugo Eulitz has survived. Of course, the road from design to construction and from testing to a functional prototype would be a long one (Fig. 12).

Carbon dioxide, or CO_2, engines work similarly to compressed air engines. Lilienthal would use a finger to pull a ring attached to a wooden handle. This opened a valve, allowing the compressed carbon dioxide to flow out of a high-pressure steel container, which actuated a piston in the engine, pulling the wing tips downward with ropes. First, Lilienthal planned to investigate the effect of single wing beats. However, he already built a repeating device into the engine so that it would be able to flap the wings multiple times when operated continuously. In June 1895, he wrote to Heinrich Bolzani in Vienna:

"My carbon dioxide engine works exactly like a steam engine. The pressure of the carbon dioxide is around 50 atmospheres at normal temperatures. I made my engines myself. They can only be used for a short period if the carbon dioxide is not heated, otherwise it will solidify. The carbon dioxide bottles that I use hold 1 kilogram of CO_2."

To heat the carbon dioxide, Lilienthal immersed the steel bottles in hot water before using them. Nevertheless, the engine always froze after just a few piston strokes during his experiments. Whenever there was evaporation, dry ice accumulated like snow.

In the spring of 1894, Lilienthal began testing on the *Fliegeberg* as planned. On May 5 of that year, Gustav Lilienthal wrote to Octave Chanute in Chicago on behalf of his brother, whose English was insufficient for the task:

> *"Our experiments with fixed wings have been discontinued, and flapping wings will be substituted. We started out with the latter type, but had to stop, because shape and equilibrium had not been sufficiently investigated. The latter difficulty has been overcome, so that we now can try flapping wings again. There is a hard road to be traveled."*

In the summer of 1894, Lilienthal took his self-developed engine out on a flight for the very first time. The engine made the aircraft very top-heavy, making it more difficult to maintain stability in flight, as well as making the landing trickier. To keep the glider withing center of gravity limits, Lilienthal mounted the heavy steel CO_2 bottles behind his position, which meant that he needed to develop long supply tubes to feed the motor.

It seems likely that extensive practice was required to master gliding with the heavier machine. The first photographs of the design in flight were published in *Prometheus* in the autumn, but, at that point, the engine was not installed.

In early August 1895, Lilienthal told Langley, who was visiting Berlin from Washington, D.C., that the carbon dioxide engine was nearly finished. He was likely referring to a new design by the young engineer Paul Schauer, who had been working at Lilienthal's factory since 1894. Lilienthal showed Langley one of the two steel bottles that he planned to carry with him on the flight. Langley estimated that the bottle contained around 2.5 liter (0.7 gal.) of gas and weighed 3.5 kg. Lilienthal told him that this would provide enough fuel for around 100 wing beats, or two minutes of operation, and rated the engine's output at around two horsepower. Pressurized bottles containing compressed carbon dioxide were readily available commercially at this point.

Langley, however, was not convinced by Lilienthal's design, particularly the use of willow branches. He was likely comparing them against his own designs in his mind. In a letter to his employee Augustus M. Herring, he described the flying machine as follows:

> *"It was made of sticks of pollard willow. It is very crooked and irregular wood. The linen was stretched as tight as a drumhead. I could not describe the curves as I had no opportunity of measuring, but the aspect of the whole was heavy and clumsy. – However, 'Handsome is as handsome does.'"*

In the autumn of 1895, Lilienthal admitted to Wolfmüller in Munich that he had still not made much progress with the wing flapping mechanism. He wrote:

"I have not finished the moving apparatus, and I fear that I will not be able to test it before winter. My past experiments in winter have always needed to be adjourned, as clammy fingers prevent effective practice, and excessive moisture deteriorates the equipment."

Lilienthal considered that he had achieved a sufficient mastery of gliding flight. He also believed that he had reached the limits of his measuring techniques for optimizing airfoils. Conceivably, he could have equipped a large number of small models with different airfoils and evaluated their performance statistically. Lilienthal saw two ways to improve his flying technique: mastering stronger winds and incorporating wing flapping into his designs. He had already mastered wind speeds of 7 m/s. At 10 m/s, he believed that he would be able to achieve indefinite flight times without wing flapping. Another path that he considered was to increase the lift by using larger wing surfaces, but those would be more difficult to control. In 1895, he also began to experiment with alternative control methods but remained convinced that the key to longer flights had to come from successfully imitating the way that birds flap their wings. Wing flapping remained a primary focus for Lilienthal until his death and was likely the object of his final elaborate design, one which would remain untested. Unlike his other designs, his wing flapping machine remained in active development for years after 1893, equipped with various engines, until he ultimately designed and built a completely new ornithopter in 1896.

In the spring of 1896, he tested a new two-cylinder engine that Schauer had designed for the first flapping device. According to Schauer's notes, this engine was far more suitable as an aircraft engine than Lilienthal's first attempt. It was reported to weigh only 5 kg, with the pressurized bottle itself accounting for 3.5 kg. One steel bottle was able to carry 3 L of carbon dioxide as fuel for the flight, which Schauer claimed resulted in an operating time of four minutes. The two-cylinder engine was built to be operated by a lever on the wooden handle. The evaporating carbon dioxide would flow into the cylinder, driving the piston rods inward and pulling the wing tips downward via the angle levers at the base. When the carbon dioxide supply was cut off, the valves opened, and the wings and pistons returned to their initial positions. This allowed the frequency and amplitude of the wing beats to be controlled as desired, meaning that their effect could be studied.

By the spring of 1896, Lilienthal was already testing the new engine in flight. Schauer reported that he first made a few light wing beats, followed by stronger ones. "*One could clearly recognize how the wing beats had a lifting and forward propelling effect*," he added.

As convincing as it may sound, the claim needs to be viewed in context, with some skepticism. It was made by Schauer himself, who was involved in designing the engine, and not until some three decades later when Schauer was in the midst of a journalistic feud, arguing that Lilienthal had successfully made a powered flight seven years before the Wright brothers. Schauer failed to convince anyone, and there are no contemporary accounts to back up his claim, including none from Lilienthal.

Despite being well-documented, Lilienthal's first ornithopter was also lost to time. In 1900, it was acquired by the Viennese meteorologist and flight researcher Raimund Nimführ in Berlin, most likely without an engine. A brief note can be found in a book by Nimführ from 1909: "*Several years ago, Lilienthal's motorized flyer became the property of Major General Neureuther in Munich, who then dedicated it to the Deutsches Museum.*" Nimführ adds: "*The experiments performed with Lilienthal's original device showed that the material was no longer sufficiently reliable. The willow branches of the stiffening frame had already become frail, and the covering was badly damaged*."

Regarding the transfer of ownership of the flying machine to the *Deutsches Museum*, there is indeed an entry in the old registry dated October 1906 stating that a Lilienthalian aircraft originally owned by R. Nimführ and subsequently donated by Major General Neureuther would soon be placed on exhibit. However, it appears to have been in poor condition. Given that the museum's inventory already included one of Lilienthal's gliders, it seems likely that it was discarded.

In the summer of 1896, a second, much larger ornithopter design was almost completed, as reported by Wood, who visited Lilienthal on Saturday, August 1, in his factory. He reported:

"*A small corner was given over wholly to the 'Flug Apparat,' and here I found a number of men at work upon a new pair of enormous wings of more than twenty-five square yards superficial area. Of this machine he had great hopes, and explained every detail of its construction, little realizing that he was destined never to put it into actual test.*"

There can be no doubt that he was talking about a new, larger ornithopter. This new device was characterized by the fact that, when the wings flapped, the outer surface parts were no longer pulled down elastically; instead, they were hinged to the immobile inner section of the wing. There were again six

Fig. 13 Model of the *Large Wing Flapping Apparatus* in the exhibition of the Otto-Lilienthal-Museum. *Photo* Maciejewski

flight *feathers* on each side, attached to a flatly truncated wing with pivoting hinges. Two factors likely prompted Lilienthal to deviate from the elastic wing flapping design and overcome his concerns about incorporating a joint into the wing. First, he wanted to reduce the stress on the material caused by elastic wing flapping, and second, he wanted to increase the amplitude of movements at the wing tips (Figs. 13 and 14).

Lilienthal's second brand-new flying machine followed the same tried-and-tested design principle as the first. A more stable wing leading edge was attached, similar to a structure that Lilienthal had patented the previous year. On each wing, there was an additional small tensioning block, positioned precisely at the flapping joint. This additional strut protruded both upward and downward. A tensioning wire ran over its upper section, with an elastic band connecting to the main tensioning block at the hinge pocket. Pull wires for the downstroke of the wing ran over the lower section. However, these pull wires did not run diagonally down to the base point, like in the first ornithopter. Instead, they ran almost horizontally to the pull rods of the two-cylinder CO_2 engine.

The cylinders in Schauer's engine had been rearranged for installation in the large flying machine. The cylinders were now aligned with one another in front of the cross frame and attached together with a hinge at their heads.

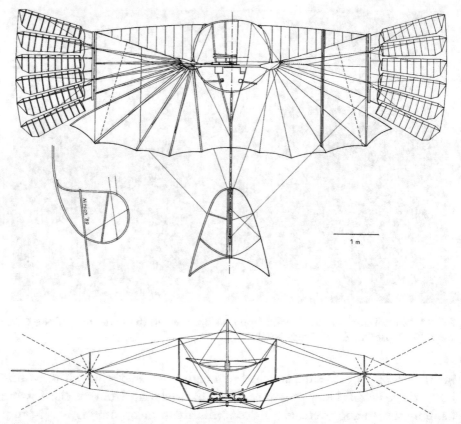

Fig. 14 Reconstruction drawing of the model *Large Wing Flapping Apparatus.* *Drawing* Nitsch. © Otto-Lilienthal-Museum.

According to Wood, the engine had not yet been installed one week before Lilienthal's death. And we have reason to believe that the gliding flight experiments without a motor had also not yet been completed. Lilienthal would undoubtedly have proceeded step by step, as always. Given the technological developments of the last 120 years, it seems doubtful that Lilienthal's wing-flapping design would have ever come to fruition. But the idea of imitating the wing flapping of birds using technology has never lost its appeal, even in modern times. Many natural scientists, aviation pioneers, universities, and bionics researchers continue to experiment. In 2011, the *Bionic Learning Network*, part of German industrial giant *Festo*, introduced their autonomous *SmartBird*, an artificial herring gull with a two-meter wingspan. In a series of impressive videos that show the *SmartBird* in flight, even in moderate wind, it can hardly be distinguished from the real thing. The company's latest

Fig. 15 The *SmartBird*, artificial gull by Festo, presented at the Hanover Fair in 2011. Festo AG & Co. KG © CC BY 3.0 (https://creativecommons.org)

design, the smaller *BionicSwift*, is perhaps even more remarkable given the lifelike complexity of the construction and movement of its wings (Fig. 15).

Decades after Lilienthal's death, Igo Etrich, a deservedly renowned Austrian aviation pioneer, caused great confusion with a statement about the whereabouts of the large ornithopter. Etrich had acquired the *Sturmflügelapparat* from Lilienthal's estate in Berlin in 1898. In 1960, at a very old age, he wrote about his efforts at the time: "The device carrying Lilienthal when he crashed had already been burned, but a duplicate and a winged flying device still existed. I bought both of them and sent them to Trautenau." He added that he later gave the small monoplane to the Technical Museum in Vienna and the wing flapping version to the *Deutsches Museum* in Munich. The latter is thought to have subsequently burned at an exhibition in Brussels (Fig. 16).

The image with two of Lilienthal's devices was published in a Viennese newspaper in 1908. The accompanying text stated:

"The workshop in Oberaltstadt near Trautenau. The flying machine under the room's ceiling formerly belonged to the aviation pioneer Lilienthal, who was killed in a crash. This same device carried Lilienthal to his death. In the background below, another one of Lilienthal's devices, equipped with a carbon dioxide engine. In the foreground, the large Wels-Etrich model flyer with a motorized engine."

But the caption is in error—the glider hanging from the ceiling is actually the *Sturmflügelapparat*, which is currently preserved at the museum in

Fig. 16 The large ornithopter and the *Sturmflügelapparat* at a point where it was owned by Igo Etrich in Trautenau

Vienna. The large ornithopter is on the ground, and the engine appears to have already been installed.

A Lilienthal ornithopter owned by the Reichau and Schilling patent bureau, which had acquired several flying machines, was presented at the Berlin International Sports Exhibition in 1906 and 1907 as the *Large Wing Flapping Apparatus*, as well as in 1907 with an appeal for donations to fund the restoration and preservation of the aircraft. The ornithopter was again shown with a carbon dioxide engine. It has not been definitively established whether this was indeed the new, larger design, and neither has survived.

Lilienthal returned to laboratory experiments, undertaking a new round of basic research into wing flapping. Aeronautical sketches that have survived in the collection of the *Deutsches Museum* include a set of very detailed colored workshop drawings for a *Rotating Experimental Apparatus with Cycloid Motion*. The drawing showing a general view is dated January 29, 1895. It shows another whirling arm device, an aerodynamic balance for measuring lift. However, this device has four test surfaces that perform flapping motions on a circular path with a diameter of 5.5 m, controlled with a low-friction chain drive running over wooden sprockets. There are two independent drives to control the flapping speed and the forward speed of the surfaces. The balance is used to measure the lift as in Lilienthal's earlier experiments (Fig. 17).

A few weeks before his death, Lilienthal announced a new series of decisive experiments in a public lecture held at the Berlin trade fair, which he claimed would represent a major step forward in the development of flight technology. Was he talking about wing flapping? It seems entirely plausible.

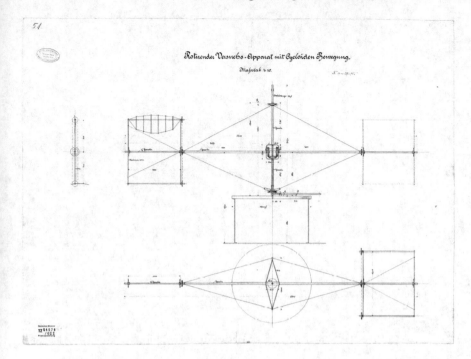

Fig. 17 Workshop drawing of the *Rotating Experimental Apparatus with Cycloid Motion*. © Deutsches Museum. All Rights Reserved

The Fliegeberg—A Real Airfield

Early in the summer of 1894, Otto Lilienthal commissioned the construction his 15-m high *Fliegeberg* near the brickworks of the Lichterfelde building association, barely two kilometers from his home as the crow flies, to serve as a permanent training location for his flight attempts. Oskar Otto, the director of the building association, had kindly given him permission to reshape a slag heap, the so-called Töpferberg, into a conical shape for this project. And so, one of Lilienthal's deepest wishes came true. The free-standing hill was evenly sloped on all sides, allowing Lilienthal to perform jumps regardless of the direction of the wind, except southeast, where the brickworks' open clay pit and two long low houses occupied by brick workers were found.

At the top of the *Fliegeberg*, whose circumference was around 200 m at its base, Lilienthal built a windowless shed where multiple flying machines could be stored after their wings were folded. Without a ledge like on the Maihöhe, which obstructed the wind and was therefore very troublesome, the shed aligned almost seamlessly with the hilltop. The shed was accessed via a beaten path form the southeast, which could not be used for flights anyway. The roof was covered with grass, allowing Lilienthal to take a few steps before jumping. At the very top, there was a stone slab at the edge, on which Lilienthal would rest for balance, always pausing for a few seconds to concentrate and wait for a favorable wind before taking off (Fig. 1).

Lilienthal's flight engineer and assistant Paul Beylich helped to build the hill. Beylich lived with his father, who was the building association's smith, in one of the two houses near the hill. He was present whenever the weather was

M. Raffel and B. Lukasch, *The Flying Man*, Springer Biographies, https://doi.org/10.1007/978-3-030-95033-0_13

Fig. 1 Lilienthal, ready to jump at the top of the *Fliegeberg*. *Photo* Anschütz, August 16, 1894. © Otto-Lilienthal-Museum. All Rights Reserved

favorable for flying exercises. "I was a carpenter, a tailor, I did everything," he said in an interview. Paul Beylich outlived his employer by almost 70 years, dying in 1965 at the age of 90. By then, Lilienthal had become famous, and Beylich was regularly courted by radio and television stations as his employee and eyewitness. He was able to fill in many of the details about Lilienthal and made full-scale models and replicas of Lilienthal's flying machines (Figs. 2, 3 and 4).

But Beylich's eyewitness accounts are problematic for a reason that will be familiar to any historian. The memories of contemporary witnesses of frequently recounted events that transpired decades ago tend to shift gradually, completely unintentionally and through no fault of their own.

Several numbers have been suggested for the cost of building the hill, but none can be verified. The only clue that we have is that Gustav Lilienthal repaid his brother a loan of 2000 marks at the time. It seems likely that this amount alone would not have covered the costs, and 3000 gold marks might be a more realistic estimate. Beylich told stories about how he carried sand up the hill with a bricklayer's trough, while other reports describe a horse-powered elevator.

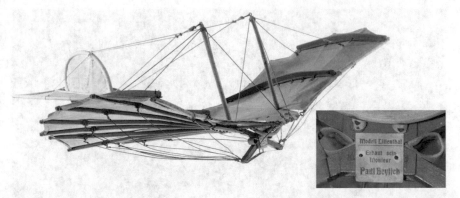

Fig. 2 Model made by Paul Beylich with a plate on the crossframe: *"Lilienthal model. Built by his technician, Paul Beylich."* © Otto-Lilienthal-Museum. All Rights Reserved

Fig. 3 Paul Beylich with models of Lilienthal's aircraft. © Otto-Lilienthal-Museum. All Rights Reserved

Once the construction of the *Fliegeberg* was complete, Lilienthal was finally free to spend every spare hour practicing flight. Not far from home, he could now begin early on Sunday mornings when the weather was suitable, taking a lunch break and then flying again until dusk, with no travel required. During the summer, he often also practiced on weekdays, starting in the late

Fig. 4 Paul Beylich with a faithful replica, around 1930. © Otto-Lilienthal-Museum.

afternoon. Word soon reached Berlin about Lichterfelde's new attraction for visitors. Anna Lilienthal, Gustav's wife, reported:

> "*Berliners have a certain instinct for this type of thing, and word soon got around that something completely new was happening here. The Fliegeberg became a popular destination for trips. On Sundays, they brought their children and camped at the foot of the hill. The 'flying man' was greeted loudly whenever he appeared and met with applause or heckling after each departure, depending on the length or brevity of the flight.*"

The first photographs of the *Fliegeberg* were taken by Anschütz on August 16, 1894. They were published in *Prometheus* the very same year, showing Lilienthal in flight with his new standard monoplane, the *Normalsegelapparat*. From this point onward, Lilienthal tested all of his designs at the *Fliegeberg*, including modifications to existing flying machines, demonstrating them to visitors from around the globe.

From a height of 15 m, including a few for the approach, Lilienthal could fly distances of 50–60 m from the *Fliegeberg*, or even up to 80 m under ideal conditions. This was not much compared to the flight distances that he achieved in the Rhinower Mountains but was plenty to test his innovations.

Fig. 5 At the *Fliegeberg* with a large audience. © Otto-Lilienthal-Museum. All Rights Reserved

Even so, Lilienthal expressed the desire to build a hill that would be twice as high near Berlin, from which he would be able to achieve approximately the same flight distances as he had done in the countryside in Rhinow.

In 1895, Lilienthal wrote in *Prometheus* that a sports field allowing young men to practice gliding would be an excellent attraction for both the participants themselves and any curious onlookers. If a real flying competition could be held from time to time, popular festivals like those surrounding other sporting competitions would undoubtedly soon develop:

> "*One can see even now that the pleasure and interest of the public in such races, when the gymnasts skilled in flights, shoot through the air, would be greater and more intense than, for instance, in horse or boat racing.*"

What Lilienthal needed was a suitable starting place. He wrote:

> "*The only thing which may cause difficulties is the procuring of a suitable place for practicing.*"

Fig. 6 Besides the large audience, a second photographer is visible on the right of the picture, taken on June 29, 1895. Dr. Richard Neuhauss was taking photographs on this day, and so was Alex Krajewsky. Some of the photographs in Lilienthal's estate cannot be clearly attributed to their photographer. © Otto-Lilienthal-Museum. All Rights Reserved

"Just as the starting from the earth is rather difficult for larger birds, the human body, being still heavier, meets with peculiar difficulties at the first flight upward. The larger birds take a running start against the wind or throw themselves into the air from elevated points, in order to obtain free use of their pinions. As soon, however, as they float in the air, their flight, which was begun under special difficulties, is easily continued. The case is similar in human flight. The principal difficulty is the launching into the air, and that will always necessitate special preparations. A man will also have to take a running start against the wind with his flying apparatus, but on a horizontal surface even that will not be sufficient to free himself from the earth. But, on taking a running start from a correspondingly inclined surface, it is easy to begin one's flight even if there is no wind.

According to the example of birds, man will have to start against the wind; but as an inclined surface is necessary for this, he needs a hill having the shape of a flat cone, from the top of which he may take starts against the wind in any direction. Such a place is absolutely necessary, if one wishes to make flying experiments in a convenient way without being dependent on the direction of the wind.

For this purpose, I have had an artificial hill, 15 meters high, erected near my house in Gross Lichterfelde, near Berlin, and so have been enabled to make numerous

Fig. 7 At the foot of the *Fliegeberg*. The factory buildings of the brickworks are visible in the background. Lilienthal is flying with an enlarged vertical stabilizer. *Photo* Krajewsky, 1895. © Otto-Lilienthal-Museum. All Rights Reserved

experiments. The drawings show this hill, or part of the same, from the outside. [The figure] represents a section of it, showing the cavity in the top intended for keeping the apparatus. At the same time the line of flight taken in calm weather is shown by dotted lines.

If a place for this sport is procured where young persons wishing to indulge in flight can disport themselves in the air, they will then have a chance to make instructive and interesting sailing flights, and I should advise having the hill twice as high, and to form it according to [the second figure], so that one can commence the flights from a height of 30 meters. The cavity inside should be large enough to hold several complete machines.

From such a hill one can take flight of 200 meters distance, and the floating through the air on such long distances affords indescribable pleasure. Added to which this highly exciting exercise is not dangerous, as one can effect a safe landing at any time.

Such a place in which young men can practice sailing flights and can at times make motor experiments with the wings would prove to be of great interest, both to those participating and to the public in general.

And when, from time to time, competitive flights were arranged, we should soon have a national amusement in this as in other sports which we have already. [...] The air is the freest element; it admits of the most unfettered movement, and the

Fig. 8 Lilienthal in flight. On the ground, his wife Agnes Lilienthal and Paul Beylich. *Photo* Krajewsky, 1894. © Otto-Lilienthal-Museum.

motion through it affords the greatest delight not only to the person flying, but also to those looking on. It is with astonishment and admiration that we follow the air gymnast swinging himself from trapeze to trapeze; but what are these tiny springs as compared to the powerful bound which the sailor in the air is able to take from the top of the hill, and which carries him over the ground for hundreds of yards?

If the atmosphere is undisturbed, the experimenter sails with uniform speed; as soon, however, as even a slight breeze springs up, the course of the flight becomes irregular, as indicated in [the figure]. The apparatus inclines now to the right, now to the left.

The person flying ascends from the usual line of flight, and, borne by the wind, suddenly remains floating at a point high up in the air; the on-lookers hold their breath; all at once cheers are heard, the sailor proceeds and glides amid the joyful exclamations of the multitude in a graceful curve back again to the earth.

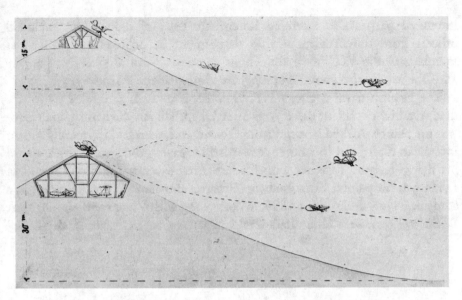

Fig. 9 The actual *Fliegeberg* compared to Lilienthal's ideal hill, which would have been twice as high. Such a hill, he wrote, could serve as a sports field for young flight enthusiasts. © Otto-Lilienthal-Museum. All Rights Reserved

Can any sport be more exciting than flying? Strength and adroitness, courage and decision, can nowhere gain such triumphs as in these gigantic bounds into the air, when the gymnast safely steers his soaring machine house-high over the heads of the spectators.

That the danger here is easily avoided when one practices in a reasonable way, I have sufficiently proved, as I myself have made thousands of experiments within the last five years, and have had no accidents whatever, a few scratches excepted.

But all this is only a means to the end; our aim remains — the developing of human flight to as high a standard as possible. If we can succeed in enticing to the hill the young men who today make use of the bicycle and the boat to strengthen their nerves and muscle, so that, borne by their wings, they may glide through the air, we shall then have directed the development of human flight into a course which leads toward perfection."

For Lilienthal, the major question that remained was who would be willing to finance such a hill near the capital. He claimed that suitable grounds had already been made available by a rich benefactor for the next few decades, but he needed a second benefactor to complete the construction and create a place where young people could learn to fly and develop aviation into a real sport once and for all.

Between 1894 and 1896, the year that Lilienthal died, curious onlookers visited the *Fliegeberg* in droves, but they were not alone—many seriously interested parties also attended Lilienthal's feats of flight. This included officers from the Prussian airship department in Schöneberg, as well as members of the VFL in Berlin. Flight engineers from abroad like Langley and Zhukovsky also counted among the visitors at the *Fliegeberg*.

The English naval engineer Percy Pilcher came to Lichterfelde twice. He died in a flight accident three years after Lilienthal. Another intrigued visitor was the painter Arnold Böcklin from Switzerland, who had repeatedly shown interest in flight, but had never succeeded in leaving the ground with any of his flying designs. Together with his father, who was a childhood friend of Lilienthal, the painter Ernst Ludwig Kirchner also came to the *Fliegeberg* as a student. In July and August 1896, Wood, along with Viennese flight engineer Wilhelm Kress, were Lilienthal's final visitors.

Production, Sales, and Imitations

Lilienthal considered his *Normalsegelapparat* with a 13 m² wing to represent a kind of standard model, as suggested by his choice of name (Fig. 1). According to Beylich, 12 copies of this model were built, and nine of the buyers' names are known. Despite the low number, Lilienthal's *Normalsegelapparat* gliders were sent to customers around the world. Lilienthal's price of 500 marks for the glider was, as noted, a considerable sum, a fact that likely contributed to the lack of sales to young people that he'd hoped for. However, Lilienthal never tired of promoting the prospective sport of flying in lectures and essays. On June 21, 1895, he gave a lecture at the *General Exhibition for Sport, Games, and Gymnastics* entitled: *The Art of Flying as a Branch of Gymnastics*. An introductory lecture had been given earlier by the renowned social ethicist Moritz von Egidy. His lecture, *Gymnastics, Games, and Sports are Elements of Popular Education,* was printed in the exhibition catalogue, together with Lilienthal's lecture. Lilienthal explained:

"Only the interplay of three factors: 1. scientific foundations, 2. technical implementation, and 3. gymnastic practice, can offer us secure prospects of progress in our solution of the problem of flight. Neglecting any one of these three factors makes successful education in the two others impossible. Thus, the lack of engagement with gymnastic flight exercises has hindered the development of a purely scientific treatment of flight, as well as the constructive perfecting of the problem of flight."

M. Raffel and B. Lukasch, *The Flying Man*, Springer Biographies, https://doi.org/10.1007/978-3-030-95033-0_14

Fig. 1 Otto Lilienthal's own *Normalsegelapparat* after his deadly crash, in the yard of his factory 1896. *Photo* Regis, 1896. © *Otto-Lilienthal-Museum. All Rights Reserved*

With increasing popularity, Lilienthal found more success in America. In response to an advertisement in *Moedebeck's Pocketbook for Flight Technicians and Aeronauts*, the American flight engineer Octave Chanute asked on behalf of Capt. William Glassford of the U.S. Army whether Lilienthal would be prepared to deliver a glider with a complete set of instructions to the United States.

An aviation journalist from Boston named James Means proposed opening an actual flight school together with Lilienthal. In 1895, Means began publishing the most important U.S. aviation periodical after Chanute's celebrated series of articles *Aeronautics* titled *The Aeronautical Annual*. He had previously published a short book called *The Problem of Manflight* in 1894. Otto Lilienthal's flying machine was illustrated on the cover (Fig. 2).

Means agreed with Lilienthal's approach, using the appeal of flying as a sport to drive the field forward:

"Have regattas and large prizes. Appeal to the people's love of sport and show what possibilities of recreation have been suggested by the experiments of Otto Lilienthal. Tobogganing on ice we can have only a few weeks in the year: tobogganing on air is possible at all seasons. When we have made our aeroplanes or aerocurves automatic in their steering action, flights like Lilienthal's will be, to say the least,

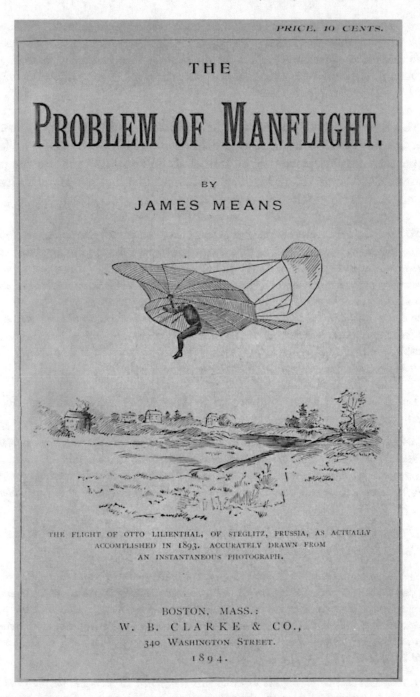

PRICE, 10 CENTS.

THE

PROBLEM OF MANFLIGHT.

BY

JAMES MEANS

THE FLIGHT OF OTTO LILIENTHAL, OF STEGLITZ, PRUSSIA, AS ACTUALLY
ACCOMPLISHED IN 1893. ACCURATELY DRAWN FROM
AN INSTANTANEOUS PHOTOGRAPH.

BOSTON, MASS.:
W. B. CLARKE & CO.,
340 WASHINGTON STREET.
1894.

Fig. 2 Cover of James Means' *The Problem of Manflight*, Boston 1894

no more dangerous than football and quite as interesting. In order to encourage the designing and construction of soaring machines, I suggest that a sum of money be raised to be offered as a prize to the constructor of the most successful soaring-machine, the award to be made after a public trial of the same, to take place early in September of the present year"

Means published Lilienthal's article *Practical Experiments for the Development of Human Flight* in the second volume of *The Aeronautical Annual* and requested a detailed account of all results of the upcoming year for the next volume. On March 10, 1896, Means offered Lilienthal the opportunity to open an actual flight school in America:

"What is needed now is that you should come over here to America for two or three months and let people see you in the air and hear from your own lips the statement that other athletes can learn the art which you have acquired. I think that would awaken an immense amount of interest here and hasten the finale solution of the problem. I have been talking lately with some of my aeronautical friends in regard to this matter and we think it is best to ask you whether you would consider the matter of coming over here to the United States in the autumn of this year, say, about the 15th of September. If so, what sum of money would it be necessary for us to pay you if you were absent from Berlin about two months?"

Ten days later, this was followed with a concrete offer of a contract:

"We want to make a binding contract with you which shall cover an absence from Berlin of not more than three months and not less than two months. We want to know how much money it will take to do this. We want you to visit this country and bring some of your machines with you. The number of machines is to be decided later.

1. Object of your Visit: The object of your visit is to introduce to the people of the United States the use of the air-sailing machine as a SPORT.

2. Expenses If the contract is made all of your travelling expenses and living expenses from the time of your leaving Berlin to the time of reaching it again, are to be paid by ourselves. The expenses connected with your flying equipment, that is to say, its construction, repairs and transportation are to be paid by yourself. The expenses of a starting hill or a revolving slope, (as described in the 1896 Annual, p. 102) are to be paid by us.

3. Premium Besides paying the expenses just mentioned in Sect. 11, we are to pay you for each month of your absence from Berlin a certain number of pounds sterling as a premium. If we make a contract with you, Mr. Millet and I should ask you during the term of that contract, to agree not to exhibit any part of your apparatus or furnish any information to anyone without our consent. Inasmuch as we wish

*to raise a fund to be devoted in the future to the cause of aeronautical science, we
must be free to make such business arrangements as may be necessary."*

This was followed by questions about Lilienthal's wishes regarding the
flight hill that should be constructed and some shorthand code words to make
it easy to reply by telegram to arrange the desired dates. Lilienthal ultimately
declined the opportunity. He was not just the owner of his factory, he was
also its manager, chief designer, and project developer. In a letter dated April
17, 1896, he offered an apology:

*"Unfortunately, my local business does not afford me the opportunity to absent
myself from it for extended periods of time. Large jobs and the great Berlin trade
fair compel me to remain here this year. I would very much like to arouse the
interest of the American people for the sport of flying by travelling to America, and
I am firmly convinced that many disciples for this art would be found if I had the
opportunity to demonstrate my experiments there."*

Means replied on April 30, expressing his understanding and hopes that
the project could be realized in 1897.

*"Will you please send to me promptly everything that is printed concerning your
1896 experiments?"*

Three articles by and about Lilienthal were published posthumously in *The
Aeronautical Annual* of 1897.

The *Normalsegelapparat*, the world's first production aircraft, perfectly
met Lilienthal's objective of being marketable as sports equipment. It was
collapsible, easy to transport, straightforward to assemble, use and repair—a
piece of recreational gear that anybody could use. It had a wingspan of 6.7 m
and a maximum chord of 2.4 m. Each wing consisted of nine ribs and was
only covered on the upper side, like previous designs. Two profile rails were
inserted into the wings on each side. The profiles were safer, now less curved,
with the additional arm pads that Lilienthal had added on the frame circle
to support the upper arms from behind. The horizontal tail surface had been
moved right to the back. Together with the vertical fin, it formed the empen-
nage—or *tail*, as Lilienthal called it—characterized by the crosswise stabilizer
assembly.

By 1893, Lilienthal was already in discussions for multiple orders. The first
known delivery was made in March 1894 to Charles Brown, a Swiss machine
designer. Heinrich Seiler from Germany was another customer in the summer
of 1894; he received instruction in person at the *Fliegeberg*. In the summer

of the same year, a glider was dispatched to Count de Lambert at Versailles in France.

Another glider is currently preserved in the *London Science Museum*. Today, it is difficult to determine the original condition of the surviving aircraft. Willow branches and canvas are not materials designed to stand the test of time. All original devices have now been restored several times or replaced with replicas. In the 2010s and 2020s, the *Deutsches Museum* in Munich and the *National Air and Space Museum* initiated a series of elaborate projects to preserve the original remains of the gliders and display them to visitors as "relics of aviation history" alongside complete replicas.

Lilienthal provided his buyers with detailed instructions on how to set up and use the gliders. He was also very liberal with his advice for flight training. One year before his own flight accident, he wrote to his customer Wolfmüller in Munich (Figs. 3 and 4):

"Practice always plays a major role. If you neglect to practice, you risk paying a steep price. Being tossed back and forth in the air with no ground under your feet is truly no fun at all. For my own part, I consider my greatest technical accomplishment in flying to be that I have never broken a bone in my attempts. There have been contortions, sprains, and wounds aplenty, but they do not count, as they do not prevent work for long. Never forget that you only have one neck to break."

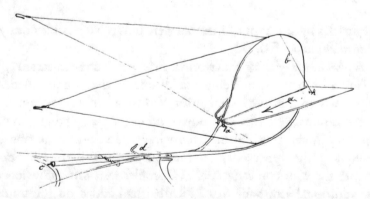

Fig. 3 From instructions written by Lilienthal for Alois Wolfmüller on company paper from his factory, dated December 13, 1894. The hand drawing shows the suspended horizontal stabilization surfaces on the Normalsegelapparat. The horizontal stabilizer and its slot are placed over the vertical stabilizer, connected with pin "a". The tether "b" places it at a negative angle, so that the back end is slightly higher than the pivot point in front. The crosswise stabilizer assembly is connected to the flying apparatus with pins "c" and "d"

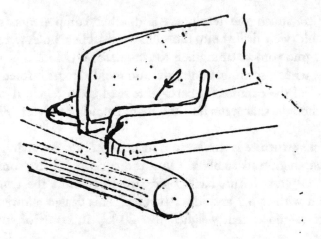

Fig. 4 The profile rails are inserted from the front through iron lugs on each rib and fixed to the front edge of the wing with peg "f". The rails are shaped like an upside-down "T". They give the bat-shaped wing the desired curvature. Since they are easy to exchange, Lilienthal could test different profile shapes consecutively on the same device. Today, similar components are found on modern planes, called the "wing fence". They reduce undesirable crossflow of air over the wing

The drawing of the standard model also clearly shows that the tail was now a separate structural unit that could easily be replaced. This means that photographs showing different tails do not necessarily imply different models. This applies in particular to a visibly enlarged vertical stabilizer that Lilienthal tested on several of his flying models.

Detailed instructions for setting up the monoplane were included in the package for the customer. After the wing surfaces had been unfolded, hooks on the front pull wires needed to be inserted into eyelets on the front edge of the frame ring. This tensioned the device, requiring some force, so a cord was attached to the center of the frame ring to assist with the tensioning. The tensioning braces or "struts" were then attached to the hinge pockets. The entire structure was built according to simple and consistent principles. From a modern perspective, they may seem archaic, but in production they offered a perfect combination of light weight, stability, and simplicity. Lilienthal chose willow as a material not because of lack of alternatives, but after careful research:

> "*The various designs of our test objects convinced us that metals cannot be used at all to build the wings, and the wings of the future will likely consist of willow branches with a lightweight fabric covering. Bamboo canes do not adapt to the wing shapes as easily as willow wood, which grows conically but can nevertheless be worked to a certain degree without any disadvantages.*"

In order to attach the tensioners and other components, small, two-millimeter holes were drilled into the willow rods. These holes were reinforced with glued hemp cord to strengthen the connection points. The wire, hemp cord, willow, wicker, shirting, bone glue, and molded parts formed from strips of metal—all of the component parts—are used consistently, demonstrating the careful thought that went into the manufacturing process (Figs. 5, 6, 7, 8 and 9).

The device's structure is similar to a suspension bridge, relying on tension to provide strength and stability. Only the cockpit, which consists of the frame ring and cross frame, has a rigid structure, while the wings and tail are fixed to it with upper and lower bracing. This design allowed Lilienthal to manufacture gliders that weighed just 20 kg. In terms of structure and

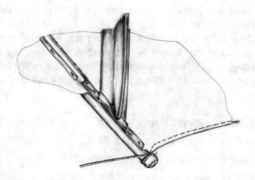

Fig. 5 The rear edge of the profile rails is held by a metal clamp. A secure connection between the fabric and the willow wood is provided by two strips of wicker nailed on with a separating gap. © Otto-Lilienthal-Museum. All Rights Reserved

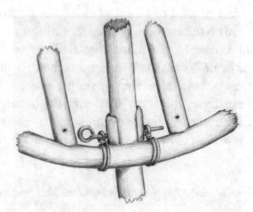

Fig. 6 Movable connection between the vertical and horizontal stabilizers. © Otto-Lilienthal-Museum. All Rights Reserved

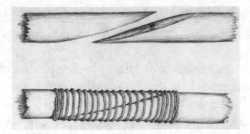

Fig. 7 Willow connection via mounting. © Otto-Lilienthal-Museum. All Rights Reserved

Fig. 8 Bracing stop with a turnbuckle on the willow rod. © Otto-Lilienthal-Museum. All Rights Reserved

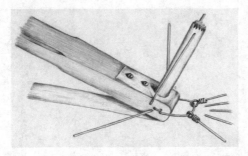

Fig. 9 Mounting of the bracing on the cross frame. © Otto-Lilienthal-Museum. All Rights Reserved

weight, Lilienthal's devices are strikingly similar to the early aluminum and Dacron hang gliders of the 1960s and 1970s.

A turnbuckle designed by Lilienthal provided the needed tension to the wires. Several contemporary accounts report that the covering was stretched as tightly as a drum, and that it would make a booming noise when rapped with the knuckles. In the event of bad weather, Lilienthal reported in *Prometheus* in 1895, the glider could be folded up within half a minute.

But pilots could just as easily wait for a sudden rain shower to end by taking shelter under the wings. Even heavy rain would not damage the impregnated canvas.

Lilienthal made thousands of gliding attempts from the *Fliegeberg*. Although he left no record of the number of flights, we can estimate that he made 20 or 30 flights per day in good weather. In a lecture to the VFL, he claimed that he had already flown 15 miles over the past year. When his audience paused, he added with a smile: *"Of course, not all at once, but across every flight this year combined,"* prompting a hearty laugh. Over the five years of his practical flying career, he undoubtedly performed several thousand flights. The average duration of his flights on the *Fliegeberg* was likely only 10–12 s, but glider pilots who take off with bungee cords on slopes will know that gliding flights of just 50–60 m distance are as dangerous as they are exhilarating. Lilienthal was the first to experience this duality of sensations. Right to the day he died, he never tired of praising the beauty of flight and the elation that it brought him (Figs. 10, 11, 12 and 13).

With nine *Normalsegelapparaten* now sold, there are a few obvious questions. Did this truly mark the beginning of aviation as a sport? Did the

Fig. 10 This attitude looks more dangerous than Lilienthal felt it to be. He still has enough distance from the ground and enough time to push the raised surface back into the normal position. Lilienthal commented on this photo by saying that it shows *"what tricks you need on rides like this through the air to avoid being thrown out of the saddle by the wind and also return the aircraft back to solid ground unharmed"*. *Photo* Anschütz, September 14, 1894. © Otto-Lilienthal-Museum. All Rights Reserved

Fig. 11 A gliding flight in calm weather on the *Fliegeberg*. Lilienthal's assistant Beylich is standing at the edge of the grass. The rain has carved several small gullies into the slope. On weekends, the hill was a popular destination for curious Berliners to see the *Flying Man* perform his feats. *Photo* Anschütz, 1894. © Otto-Lilienthal-Museum. All Rights Reserved

buyers ever make successful flights with their flying machines? In addition to the nine that sold, many other attempts were also made with look-alike of Lilienthal gliders reproduced from known publications. Lilienthal did not seem to object to this. Although he made efforts to sell his patent in the USA in his correspondence with Chanute and Means, he also corresponded with individuals who were attempting to fly with gliders they'd built based on Lilienthal's designs. Augustus Herring build several gliders after Lilienthal's model, making further improvements to the design, and Chanute even applied for a patent on his improvements to fit a seat into Lilienthal's glider (Fig. 14).

Chanute also designed and built an experimental glider that eventually became known as *Katydid* that he tested on Lake Michigan in 1896. A multi-wing design, this glider was built like a railway bridge, which is perhaps not surprising given Chanute's background as a civil engineer. Ferdinand Ferber once stated that flight developed in two lines: the poetic Lilienthal glider and the engineering Chanute glider. The Chanute glider is arguably a concrete link between Lilienthal's designs and those of the Wright brothers (Figs. 15 and 16).

Fig. 12 Lilienthal often practiced on weekdays at the *Fliegeberg* in Lichterfelde when the weather was favorable, in the evenings after returning from Berlin. He was always happy to demonstrate his *"art of flying"* to visitors, whether they were former classmates or members of the German VFL in Berlin. Within this association, Lilienthal represented an aviation movement that advocated the realization of human flight with heavier-than-air flying devices, as opposed to the supporters of airships. *Photo* Anschütz, August 16, 1894. © Otto-Lilienthal-Museum. All Rights Reserved

Fig. 13 Many photographs were taken of Lilienthal's flights in this position: gliding in a western direction, with a few of the houses of Lichterfelde in the background. The photograph is mounted on a representative cardboard, as was typical for Alex Krajewsky, 1895. © Otto-Lilienthal-Museum. All Rights Reserved

Fig. 14 *Lilienthal Apparatus* built by A. Herring. According to Chanute's records, the maximum flight distance was 70 ft. The glider was described as *"cranky"*, June 22, 1896. © Otto-Lilienthal-Museum. All Rights Reserved

Fig. 15 The *Chanute Biplane*—on the path to the Wright brothers' flyer design, 1896. © Otto-Lilienthal-Museum. All Rights Reserved

Fig. 16 The *Katydid*—experimental aircraft with many surfaces in various positions, Chanute, July 1, 1896. © Otto-Lilienthal-Museum. All Rights Reserved

Lilienthal's Customers

1 Charles Eugene Lancelot Brown

Brown, the buyer thought to be the first, was an engineer of English descent who later co-founded the globally renowned company *Brown, Boveri and Cie* in Baden, Switzerland. Brown was interested in all kinds of technical innovation, including the idea of flight. In 1910, he described himself as well-versed in all matters of aviation, having followed its development from the very beginning.

His son, Charles Brown Jr., reported in 1917 that his father had regularly corresponded with Lilienthal and had even visited him multiple times in Berlin. They would likely have discussed the prospect of opening a manufacturing branch for aviation within Brown's company. He allegedly tried Lilienthal's glider himself, in the shadow of Schartenfels castle, which towers above the surrounding industrial buildings and spa resort. This meeting must have ended with the two deciding that manufacturing flying machines was not a viable prospect for the time being.

Brown donated his glider to the *Deutsches Museum* early in 1905. In the accompanying letter, he wrote that the aircraft had suffered somewhat during his experiments but should nevertheless be of some interest to the museum. The *Deutsches Museum* did not display Brown's glider as part of its collection, presumably because they were already exhibiting Lilienthal's biplane. At the request of the Kaiser's government, however, the monoplane was loaned to the *International Exhibition* in Milan in 1906, then shown at the *International Airship Exhibition* (ILA) in Frankfurt am Main in 1909, followed by

M. Raffel and B. Lukasch, *The Flying Man*, Springer Biographies, https://doi.org/10.1007/978-3-030-95033-0_15

the *International Industrial and Commercial Exhibition* in Turin in 1912. In 1922, however, it was deemed no longer preservable or restorable and was discarded from the museum's inventory.

2   Heinrich Seiler

A drawing by Lilienthal entitled *Seiler's Apparat* has been preserved in the *Deutsches Museum*. The young man from Liegnitz in Silesia visited Lilienthal in the Rhinow Mountains in 1893 and, as he was lighter, is said to have even surpassed Lilienthal in flying. The device still existed in 1924 and was offered for sale. In the 1920s, Seiler became an active glider pilot on the Wasserkuppe, the gliding center in the Rhön Mountains.

3   Count Charles de Lambert

Like most of the others, the glider delivered to the Frenchman de Lambert was not preserved. The French aviation pioneer Ferdinand Ferber, a follower of Lilienthal's school of thought, later reported that de Lambert had tested the glider near Versailles but did not persevere for long and soon turned his attention to experimental hydrofoil boats instead. A decade and a half later, in October 1908, de Lambert returned to aviation and became Wilbur Wright's first student at Camp d'Anvours near Le Mans, France. One year later, he acquired French pilot's license No. 8 and circled the Eiffel tower in Paris in a Wright biplane on October 19, 1909,—a feat that many found sensational. It was the first time that an airplane had flown above the myriad houses of a city with more than a million inhabitants.

Surprisingly, there are no references whatsoever to de Lambert's Lilienthal glider, neither in the archives of the *Paris Flight Technical Association*, nor in the specialized French journal *L'Aéronaute* that was already in print at the time. *L'Aéronaute* reported several times on Lilienthal and his flying machines. Toward the end of 1894, it encouraged its readership to reproduce their own such machines and experiment with them in France. But there is nothing about de Lambert and the fact that he had already owned an original glider for three months. It also seems strange that de Lambert himself never referred back to this early chapter of his preoccupation with flight technology. The *De Lambert Model* design drawing by Lilienthal (see Fig. 12.9, p. 102) contains many details about the dimensions of the device. Unless they were adapted to de Lambert's body weight, they appear to correspond to an early design before the development of the final *Normalsegelapparat* was fully complete.

Nothing is known about the whereabouts of the glider. The Lilienthal glider presented in the *Musée de l'Air in Meudon* near Paris, today's *Musée de l'Air et de l'Espace* at Bourget Airport, is not an original but a replica from the 1920s, made by Hans Richter in Berlin. Richter, the man who portrayed Lilienthal in the 1920s film, also built several gliders to be exhibited in Berlin and at Lilienthal's birthplace of Anklam. Some of the details of these devices differ from the originals. Since Richter designed aircraft himself, the deviations presumably represent improvements based on his knowledge and experience.

4   Alois Wolfmüller

In addition to his interest in aviation, Wolfmüller, was an engineer who built motorcycles in Munich. He sent his first request for a monoplane glider to Lilienthal in the autumn of 1893. In November, Lilienthal replied that he could sell him a glider for 300 marks that would take four weeks to manufacture. Wolfmüller sent another request in September 1894, to which Lilienthal responded with an offer for a "*soaring apparatus like the one I use myself*", the *Normalsegelapparat*, for 500 marks. Advance payment of the purchase price was explicitly requested and appears to have been made. Just three months later, on December 13, 1894, Lilienthal announced that he had handed the glider over to the railway company for express delivery to Schongau, on the Lech River.

Before the end of December, Wolfmüller had already begun his own experiments at Lechfeld, fearing neither strong winds nor cold. He must have reported about his experiments to Berlin, as Lilienthal wrote to him in January 1895:

"*I am extraordinarily pleased that you are proceeding so energetically with your aviation experiments.*"

Between 1893 and 1896, the pair maintained a lively exchange of ideas. Eleven of Lilienthal's letters and two of Wolfmüller's letters are known. The letters also discuss Wolfmüller's thoughts about controlling a flying machine by mechanically twisting the wing. Lilienthal replied that he had already conducted experiments with wing warping, as well as other methods of mechanical control, but that he was not satisfied with the results so far.

After Lilienthal's death, Wolfmüller also suspended his own experiments, but resumed them in 1906 and 1907 with his own gliders. An original model is preserved in the *Deutsches Museum* in Munich. Some hints of the exchange with Lilienthal can be recognized in its design. Wolfmüller added control elements to Lilienthal's bat-like structure. The pilot sits on a board and operates the control elements with his upper body, similar to the solutions that Lilienthal was working on. Wolfmüller's surviving glider is also remarkable in that it used a sliding weight that could be moved fore and aft to adjust the center of gravity depending on the size of the pilot—in effect, a rudimentary trim system.

5   Dr. Kilian Frank

The glider delivered to Dr. Kilian Frank in Karlsbad (Karlovy Vary, a spa city in the Czech Republic, Bohemia), was given to the aviation club he'd founded. Frank had read about Lilienthal's flight experiments in *Prometheus* and reported about them to the *Karlsbad Cyclists' Association* in a lecture held early in November *1894*. His proposal to establish an aviation club that could perform its own experiments with one of Lilienthal's gliders was met with approval. Frank wrote to Lilienthal on November 4, and told him about the founding of the aviation club and asked for his support.

Lilienthal was quick to respond on November 6, expressing his delight at the founding of the club. He wrote:

> *"For solving the problem of flight, this should be regarded as a momentous event, as advancement in flight technology can only be expected through practical activities such as yours."*

Lilienthal ultimately planned a visit to Karlsbad in the next spring to teach the club members how to fly the glider. Regarding the acquisition of a glider, however, he referred Frank to the von Stach building council in Vienna, Kahlenbergbahn, which had acquired exclusive distribution rights to Lilienthal gliders in Austria.

At the time, Ritter von Stach was the chairman of the *Aviation Technology Association* in Vienna. According to the association's records, he informed the association on November 16, 1894, that he had "*arranged the installation of a Lilienthalian flying machine on the Kahlenberg*". According to the records, an artist named J. Stocklas, who was said to have visited Lilienthal in Lichterfelde, performed a demonstration with the glider. It is not clear what actually happened versus what was merely a statement of intent.

The next year, von Stach sent the glider, which had been on display for two months in the *Kahlenberg Hotel* near Vienna, to Karlsbad, where it arrived on February 13, 1895. On March 1, it was christened "*Lilienthal*" at festivities held in Weber's brewery. Lilienthal himself was informed, and he sent a reply promising to visit soon. But this visit never took place.

The first practical experiments were made in the spring of 1895 on the Horner Berg near Karlsbad, but they did not last for long. The flying machine was repeatedly damaged, and the chief engineer responsible for maintaining it, a man named Dix, was forced to replace the willow ribs. In the following spring of 1896, some members of the association returned to the Horner Berg with the restored glider. This time, the device broke as soon as it was set up, even before the first flight attempt, and it took several weeks for Dix to patch it up again.

After Lilienthal's death, the activities of the *Karlsbad Aviation Association* subsided. In 1897, only six members were left. The glider itself is thought to have still been in the possession of the owners of the Karlsabad spa resort as late as the 1930s.

6   T. J. Bennett

The glider acquired by the English dentist and aviator T. J. Bennett in Oxford in 1895 was preserved. He was interested in anything and everything to do with maritime and air travel and compiled an extensive library on these two subjects. Bennett was an active member of the *Aeronautical Society of Great Britain* (now the *Royal Aeronautical Society*) beginning in 1871. For several years, he was the assistant secretary of the group, the world's oldest aeronautical engineering association. For three decades, he maintained a lively correspondence with aviation inventors and experimenters around the globe. In 1897 or 1898, he gave part of this correspondence to Means, and it is now held by the *Library of Congress* in Washington, D.C. The collection shows that Bennett experimented with both model aircraft and model ships but does not include any correspondence with Lilienthal.

Only one letter between Bennett and Lilienthal is known, sent by Lilienthal on March 4, 1895. It confirms the dispatch of the ordered glider and contains the usual instructions. Lilienthal wrote:

> "*Your soaring apparatus was dispatched from this location on March 1 and should come into your possession accordingly.*"

The handwritten letter is two pages long. It was purchased at an auction in London in early 1986 for the *Science Museum* together with a set of contemporary photographs, one of which shows Bennett immediately after unpacking the glider. The letter contains almost exactly the same instructions for unpacking and setting up the glider as Lilienthal had previously sent to Wolfmüller. The three handmade drawings showing how to attach the tail, insert the profile rail, and attach the buffer are also identical. In the final paragraph of his letter, Lilienthal informed Bennett that he would soon be sending another device to a Prof. Fitzgerald in Dublin.

It's not clear whether Bennett ever attempted to use Lilienthal's glider himself. In 1896 or 1897, he gave it to Percy S. Pilcher. And, once again, it cannot be said with certainty whether Pilcher experimented with the glider. But he brought it with him on September 30, 1899, to a demonstration of his own glider, named *Hawk*, in Leicestershire—a demonstration that ended with Pilcher's death in a crash.

After Pilcher's death, Lilienthal's glider became the property of the *Aeronautical Society* and was moved to the *Science Museum* in London in 1920, where it was displayed for 50 years. The exhibit was retired from display in 1976 but still exists.

7   Professor George F. Fitzgerald

Prof. Fitzgerald was a physicist at *Trinity College* in Dublin, Ireland, who was very interested in to the idea of flight. He'd only just joined the *Aeronautical Society* in 1897 when he immediately became a member of the board.

On March 14, 1895, Lilienthal informed Fitzgerald that his glider had been shipped. The letter was written by an unfamiliar hand but bears Lilienthal's signature. It also includes instructions for unpacking and unfolding the well-wrapped aircraft, with the usual three drawings. This suggests that every buyer received these or similar instructions, including recommendations and advice for handling the glider and successfully completing the first flight. The end of the letter acknowledges receipt of the purchase price of 25 pounds sterling (Fig. 1).

Lilienthal also corresponded with Fitzgerald several times, even after the glider had been delivered. In April and December, he replied to Fitzgerald's letters about experiments with the glider and a proposal to connect two kites together to carry a person into the air. Lilienthal found this idea intriguing and worthy of consideration but noted that the coupled kites would be difficult to control in flight.

Fitzgerald reported about his own experiments with the *Normalsegelapparat* in October 1898 in *The Aeronautical Journal*. These experiments were conducted in the park at *Trinity College*, which was fairly flat and surrounded by tall trees, and therefore not ideal for flying. Fitzgerald began by performing the recommended standing exercises in the wind, then ran with the glider into the wind and practiced jumping from a springboard, assisted by a few students each time. He suspended the glider on strings several meters above the ground like a kite, at first unmanned, then piloted by one of his lighter students, and finally with himself as pilot.

Several good-quality photographs of these experiments have survived that clearly show details such as the buffer, the struts and bracing, and the inserted profile rods. The photographs also document the various training exercises described by Fitzgerald.

After the turn of the century, the device came into the possession of a Dr. MacCabe, who also tried it out in Dublin. It broke when one of MacCabe's friends attempted to jump off a large haystack and flipped over.

In a letter to Fitzgerald dated April 4, 1895, Lilienthal gave a sketch of how he held the handles in front of the frame circle with his hands: using an underhand grip, with the backs of his hands pointing downwards. He wrote:

Fig. 1 The *Normalsegelapparat* built by Otto Lilienthal and acquired by Fitzgerald, during a demonstration at the *University of Dublin*. © Irish Aviation Museum Dublin.

> "*I recommend once again to choose a treeless, open, sloping terrain for the exercises and to begin with small and careful jumps directly into the wind. The body needs to become familiar with the new movements gradually over time.*"

8   Nikolai Yegorovich Zhukovsky

Zhukovsky, a lecturer of mathematics at the Technical *University of Moscow* and the man who would become known as the *Father of Russian Aviation*, had previously visited Berlin as a young academic and attended lectures by the renowned physicists Herrmann von Helmholtz and Gustav Robert Kirchhoff. He read and appears to have spoken German fluently. In November 1889, after Lilienthal's book on bird flight was published, he gave a lecture on *Some thoughts about flying machines* to the *Society of Friends of the Natural Sciences* (SFNS) in Moscow. In January 1890, Zhukovsky gave a lecture on the theory of flight at a conference of Russian doctors and natural scientists. At the beginning of the twentieth century, he founded an aerodynamics laboratory in Moscow, with the first ever small wind tunnel built for aerodynamic experiments.

Zhukovsky met Lilienthal in person in the autumn of 1895 during a trip in Europe, and also visited the *Fliegeberg* and observed several of Lilienthal's biplane flights firsthand.

On November 7, 1895, he made the following statement at a meeting of the *Department of Physical Sciences* of the SFNS:

> "*The most important invention in recent years for the field of aviation is the flying machine created by the German engineer Otto Lilienthal.*"

Zhukovsky ordered his own *Normalsegelapparat*, which was delivered in June 1896 and was the last ever sold. It was later acquired by the University of Moscow (Fig. 2).

Zhukovsky's lecture was published in the first 1896 issue of *Photographic Review* in Moscow and illustrated with photographs by Preobrazhensky and a few other pictures of Lilienthal's flights. One interesting point about Zhukovksy's remarks is that they mention multiple visitors to the *Fliegeberg* from abroad. He said:

> "*When I had the pleasure of accepting Lilienthal's gracious invitation this autumn to visit him on his hill near Berlin, two other Russian technicians, a German photographer, and an American also attended the flights.*"

It is not known whether Zhukovsky and Lilienthal corresponded with one another beforehand. The reference to an invitation by Lilienthal suggests that they might have, but the invitation could also have been delivered by Preobrazhensky on an earlier visit. In any case, no letters or even any verifiable references to them have been found.

On October 27, 1896, Zhukovsky then gave another lecture in Moscow at the annual meeting of the SFNS, *On the death of the flight engineer Otto Lilienthal*, which was published in 1897 in the first issue of the magazine *Aviation and Research on the Atmosphere*. This journal was reissued by the *Aeronautical Section* of the *Imperial Russian Technical Society*. In it, Zhukovsky mentions another visit to the *Fliegeberg* following Lilienthal's death:

> "*One of aviation enthusiasts from Moscow, S. G. Lessenko, visited Lichterfelde in September 1896, after Otto Lilienthal had departed from the realm of the living. With help from Gustav Lilienthal, S. G. Lessenko visited the famous hill together with the assistant of the deceased aviator. The hill was completely neglected. After showing the flying machines, which were stored in an earth hut on the upper part of the hill, the assistant could under no circumstances be enticed to provide a flight demonstration with them.*"

This visit was implicitly confirmed by Beylich's later report that he had been asked to come to Moscow and continue Lilienthal's experiments there after the latter's death. Lessenko and Zhukovsky did not seem to be aware that Beylich had never flown himself. The fact that this note references multiple

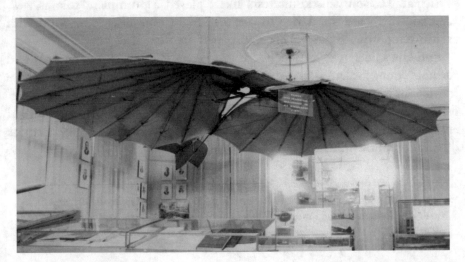

Fig. 2 The Lilienthal glider in the *Zhukovsky Museum* in the 1950s

gliders on the *Fliegeberg* is interesting in and of itself. These are presumably the ones that ultimately went to the Reichau and Schilling patent office. In the 1930s, Beylich reported that, in the factory, a few days after Lilienthal's death,

> "...*everything relating to aircraft construction had disappeared. Even a new apparatus that had taken weeks to build and been fitted with various innovations was smashed and burned in the factory boiler by the order of Gustav Lilienthal.*"

There are however photographs of the new, large ornithopter that were taken at a later date. If we are willing to trust these memories of events from more than 30 years prior, there must have been another design, a new one. There are various draft drawings and written theories that potentially corroborate this hypothesis.

The glider delivered to Zhukovsky was preserved and is now owned by the *Zhukovsky Museum* in Moscow. The museum is part of the *Central Aerohydrodynamic Institute* (TsAGI) which was founded by Zhukovsky in 1918 and which bears his name today.

9   William Randolph Hearst

Hearst acquired his glider in April 1896, apparently in response to growing public interest in the USA in aviation and in Lilienthal's experiments in particular. His journalistic interests likely played a prominent role as well. He planned to use the device to perform flight tests near New York, with exclusive reporting rights for his newspaper, the *New York Journal*.

And the glider was indeed tested by Harry Bodine, a well-known and versatile young athlete, near Bayonne, New Jersey. In a full-page article in the Journal on May 3 entitled: *A Flying Machine at last that really flies*, the glider was named *The Journal's Machine*. The article was illustrated with several engravings based on original photographs. In it, Hearst gives a very accurate description of the glider.

> "*FLYING as a sport has been tried by the Journal and found to be a success. Athletics in the air, it has been demonstrated by careful experiments, are as practicable as athletics on the ground. There is no reason why there can't be 100, 200, or even 300-yard dashes, far above the heads of spectators at Berkeley Oval this Summer. Fleetness of wing should hereafter cut quite as important a figure in the world of sport as fleetness of foot.*

Otto Lilienthal, of Berlin, who has acquired an international reputation as the 'flying man' is the inventor and maker of about the only flying machine in the world that flies. After years of careful research he perfected a soaring apparatus, constructed upon the principle of a bird's wing. It will carry a man of average weight short distances, provided a start is made from some elevation, the higher the better, and provided also the soaring is against the wind.

But this is as far as Lilienthal has got. He has not been able to build a machine that will sail great distances or in every direction. He is far from supposing that his wings, although they afford the means of sailing and even of soaring in the air, possess all the delicate and subtle qualities necessary to the perfection of the art of flight. He is satisfied, however, that his researches show that it is worthwhile to prosecute the investigations further.

With this idea in view he called upon the athletes of the world to help him. He suggested that his machine already flies well enough to test the skill and endurance of the man who operates it, and that it presented a great opportunity in athletic sports. If he could secure, he said, its general adoption for such purposes, he would be able to go on and realize its utmost possibilities. He pointed out, too, that there was nothing more fascinating than flying, while it offered infinitely more opportunities for skillful maneuvering than the bicycle.

Lilienthal is not after money. He seeks power, not profits. He only wants to have his studies of the mysteries of aerial navigation supplemented by others.

The Journal, acting upon Lilienthal's suggestion, determined to test the machine and discover if the claims of the inventor were well founded, so an order was given the German for a pair of his wings. The apparatus was constructed according to the inventor's latest model, and it arrived in this country a week ago. It covered the distance between Berlin and New York without sustaining a scratch. It resembles a gigantic sea gull. Open a mammoth umbrella and cut it squarely in half and you will have a fairly good representation of one of the flyer's wings. It is made almost entirely of closely woven muslin, washed with collodion to render it impervious to air, and stretched upon a rib frame of split willow, which has been found to be the lightest and strongest material for this purpose. Its main elements are the arched wings (Fig. 3).

In order to fully carry out Lilienthal's suggestion an all-around athlete was chosen to operate the machine. Mr. Harry B. Bodine, of Bayonne, N. J., who has acquired more than a local reputation in many kinds of sport, and who is a member of several athletic clubs was selected. He knew as much as the average man does about aeronautics and aeroplanes, which is not very much, and absolutely nothing, of course, about manipulating a flying machine. All he had to guide him were Lilienthal's special directions, which are as follows: First—Begin practice by running down a small hill, leaping high at times; face the wind directly. Second—Jump higher and higher progressively, and try to improve very slowly. Do not begin exercises while the wind is very strong. Third—When descending to earth, step out with your feet, leaning the body backward and raising the apparatus in front. Fourth—Take great care in handling apparatus at all times, especially while it is new to you."

Fig. 3 Engraving based on a known photograph. The Journal, May 3, 1896, p. 17

This was followed by a detailed account of the experiments that had been conducted, as well as an invitation to any potentially interested sports associations to participate in the tests. Spectacular progress would soon be achieved, it was announced. Two weeks later, in mid-May, another article was published, entitled *Another Successful Trip in the Journal's Flying Machine*. In August, news of Lilienthal's death reached the United States (Fig. 4).

There was no further news of the glider until it re-emerged in a badly damaged condition in January 1906 at an exhibition of the *New York Aero Club*. John Brisben Walker, editor of *Cosmopolitan* magazine, patron of aviation technology, and friend of James Means in Boston, had acquired the glider as part of the financial transaction when he sold his magazine to Hearst. Walker donated the device to the *Smithsonian Institution* in Washington in early February 1906, shortly after the exhibition closed.

Fig. 4 View of aeronautical exhibits at the *Automobile and Aero Clubs of America Joint Show* in the third-floor gymnasium of the 69th Regiment Armory, New York City, January 1906. The Lilienthal (Otto) 1894 Glider can be seen hanging on the left. © *Smithsonian Institution NASM-9A06299. All Rights Reserved*

More Difficult Than Anticipated: Control

In parallel to the production of the *Normalsegelapparat*, the experiments with new designs continued. On March 9, 1895, Lilienthal wrote to Wolfmüller in Munich:

> *"I am currently building a large soaring device with an area of 20 square meters that can only be used in the absence of wind."*

The glider in question, which Lilienthal called his "experimental device", is now known as the *Vorflügelapparat* (*front wing apparatus*) due to its distinctive structure, with slats on the leading edge of the wings. Like the second ornithopter, which was no longer being tested, it was another, significantly larger monoplane. It featured a wingspan of almost 9 m, making it nearly as big as the Südende design. The maximum chord was just less than 3 m. The profile height, i.e., camber, was 1:20 of the chord. The aircraft's large size meant that three profile rails needed to be inserted onto each wing, rather than two.

But the defining characteristic of the glider was not its size, it was the fact that various new components were added to it for the purpose of testing new control systems. A narrow, movable wing section had been added in front of the bat wing. It has become common to refer to these elements as "slats", although they have nothing in common with the take-off and landing aids of the same name that are found on modern commercial aircraft. Lilienthal thought that this front wing flap would constitute an essential safety element,

M. Raffel and B. Lukasch, *The Flying Man*, Springer Biographies,
https://doi.org/10.1007/978-3-030-95033-0_16

and he applied for a new patent to cover it. The glider also served as a test vehicle for experiments with various other control systems. On May 29 of that year, the new glider was demonstrated to a group of knowledgeable visitors on the *Fliegeberg* from the VFL. The association's brief records about this visit complain that "*the weak air flows often prevented the flyer from fully demonstrating his art*". But calm conditions were perfect for presenting the new large monoplane (Figs. 1 and 2).

The patented aspect of the device is an elongated wing strip that rotates around the front edge. The slat is folded downward on the ground, and a gap in the wing is held open with elastic bands. In flight, the incoming force from the air lifts the flap which closes the gap. The bat wing then forms a single surface with the slat. Lilienthal spoke about the dangers of "*Top Wind*" on his wings multiple times. If the device ever enters a position with too low of an angle of attack during flight and the pilot cannot shift his body weight backward quickly enough, the air flow will push the nose further downward and ultimately cause a crash. Lilienthal's leading wing flap was designed to counteract this danger by automatically opening and suddenly increasing its angle of attack by several degrees. This bought the pilot time to regain control, and once the device had been restored to a proper attitude, the air flow would cause the slat to automatically close.

Accordingly, Lilienthal described his front wing design as a soaring glider with automatic balancing. He was so convinced by his design that he applied for a patent on the same day that he performed the demonstration on the *Fliegeberg*. On August 5, 1895, he wrote to Chanute in Chicago that he had been able to significantly improve his flying machine: the new mechanical system made the device so stable in flight that anybody would be able to operate it effortlessly.

The Imperial Patent Office in Berlin issued patent no. 84417 for this innovation as a supplement to the earlier flight patent of September 1893. The patent applied to a

> "…form of the flying machine protected by patent no. 77916 in which the front part of the wing surface can be rotated downward around the front edge and pressed downward by elastic elements in such a way that it rotates downward when the air pressure ceases to act from below, thereby creating a moment that lifts the apparatus."

The drawing attached to the patent is a schematic, illustrating the patent claim without giving any details. Since the patent was issued as a supplement, the tail fin is drawn in the same shape as in 1893, even though Lilienthal had been using the crosswise stabilizer assembly for quite some time by that point.

OTTO LILIENTHAL in BERLIN.

Flugapparat.

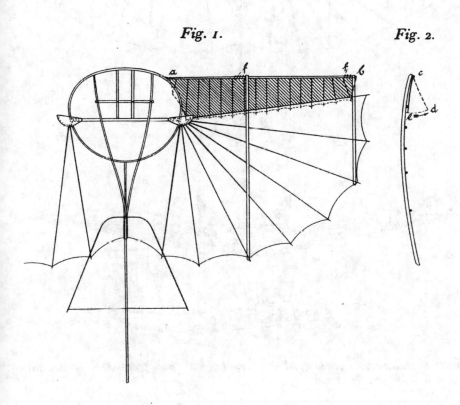

Fig. 1.

Fig. 2.

Zu der Patentschrift

№ 84417.

PHOTOGR. DRUCK DER REICHSDRUCKEREI.

Fig. 1 Illustration from Lilienthal's second flying machine patent no. 84417 (77), dated May 29, 1895. Since this patent is a supplement to the first patent from 1893, the drawing is based on the original figures of the glider from 1893. With a surface area of almost 20 square meters, the front wing glider was significantly larger than the *Normalsegelapparat*

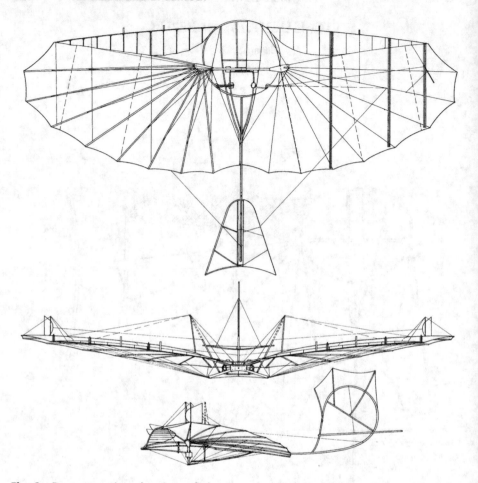

Fig. 2 Reconstruction drawing of the *Experimental Monoplane*. © Otto-Lilienthal-Museum. All Rights Reserved

On May 29, 1895, Neuhauss took photographs of the *Experimental Monoplane*. Four other photographs are known, taken by the Russian visitor Preobrazhensky. They appear to have been taken on a different day, as different control elements are installed on the glider. Small resistance surfaces can be seen at the ends of the wings, which the pilot can pivot into the air flow so that they act like lateral rudders. On Neuhauss' images, two longer control struts are attached to the device in addition to the two struts on the hinge pockets. They run from the spar cross diagonally upward through the frame circle behind Lilienthal's back, holding cords for control systems that Lilienthal was testing over the summer of 1895 (Fig. 3).

Lilienthal knew that he was already pushing the limits of feasibility with his control mechanism, which still depended on the pilot shifting his weight.

Fig. 3 Reconstruction of the *front wing glider* in the *Otto-Lilienthal-Museum.* ©
Otto-Lilienthal-Museum. All Rights Reserved

The forces that a pilot can generate by moving the legs and upper body are
not sufficient to control larger wings or even smaller ones in gusty winds.
To build a larger and more practical flying machine, mechanical controls
seemed to be a must. As already suggested by his choice of name, *Experimental Device*, he did not test the various control mechanisms operated
from the ring around his waist simultaneously. Instead, he investigated their
effects in separate experiments independently of one another on an apparatus
specifically designed for testing.

Since the *Experimental Monoplane* has not been preserved, it is possible
that it was converted into the large ornithopter in 1896. The identical
wingspans and surface areas of the two flying machines strongly suggest
that this is plausible. Furthermore, Lilienthal always used his equipment effi-
ciently, and had been known to repurpose aircraft for new applications in the
past. There is no other reliable information about the aircraft's whereabouts
(Figs. 4, 5, 6 and 7).

Lilienthal did not report on his experiments with mechanical control
mechanisms in any of his essays or lectures, only in letters to Wolfmüller
in Munich. Wolfmüller was arguably the correspondent who inspired Lilienthal the most when it came to the theories of aviation. Surprisingly, they
never met in person, although Wolfmüller had made specific plans to visit
Lichterfelde.

Fig. 4 The *"Flying Apparatus with Automatic Balancing"* was demonstrated on the same day that Lilienthal applied for a patent, May 29, 1895, on the *Fliegeberg*. Paul Beylich, who regularly assisted at the foot of the hill, is standing with the aircraft. The slat is clearly visible when folded down. It is held in place by elastic bands and closes itself during normal flight. In critical flight positions, however, the flaps open, generating a moment that causes the glider to stabilize itself. Photo: Neuhauss. © Otto-Lilienthal-Museum. All Rights Reserved

After exchanging assurances of *"accommodating reciprocity subject to the protection of legitimate interests"*, Wolfmüller presented his thoughts and the results of his experiments on controlling flying machines built after Lilienthal's design in a long letter dated September 28, 1895. This prompted Lilienthal to speak freely about his own attempts. Wolfmüller had designed a wing warping mechanism, as well as a structure that would allow pilots to sit within the flying machine. He argued that a sitting position would be advantageous, freeing up the pilot's hands to operate mechanical control systems such as the two levers used for twisting the wings in his own design. Other control elements could be operated using a strap around the upper body.

In October, Lilienthal replied:

"I tested an arrangement similar to yours for moving and rotating the wings with outer tensioning wires running to different points of a lever mounted at the lower base point that can be pulled to give the wing profile the desired rotation. I also made it so that the tail could rotate right and left, making it easier to land. Finally, I attached a surface to each wing tip that could be straightened by pulling a cord to pull back the leading wing tip. These elements were operated from the hips, which press against a sliding bar when the body is shifted sideways to regulate the center

Fig. 5 Lilienthal in flight with the *Experimental Monoplane*. In normal flight positions such as the one shown in this image, the flap is firmly closed, forming a single surface with the wing. The two additional control struts protruding backward from the frame circle are the only obvious feature identifying this machine as the front wing glider. These control struts hold cords that allow Lilienthal to operate experimental control mechanisms. Photo: Neuhauss © Otto-Lilienthal-Museum. All Rights Reserved

of gravity. But in truth I am not convinced with these innovations – if the body is free to shift the center of gravity quickly enough, the same result can ultimately be achieved by other, simpler means. As always, practice remains key.

My opinion on the question of a sitting position for the pilot is similar. I have tested various seating arrangements, but ultimately always rejected them because they do not allow sufficient freedom of motion in windy conditions. For example, one needs to be able to shift the lower body and legs quite far back to move the center of gravity far enough backward if the apparatus begins to descend too steeply. I have on occasion needed to enter the position described in order to straighten up the apparatus quickly enough."

Lilienthal concludes by admitting that he had not yet achieved a decisive breakthrough in controllability:

Fig. 6 Flights with the *Experimental Monoplane* on the *Fliegeberg* to the southwest. In the background, the chimney, boiler house, and steam brickworks are visible. Between them is the clay pit whose slag was used to build the *Fliegeberg*. Photo: Neuhauss. © Otto-Lilienthal-Museum. All Rights Reserved

> "*These experiments, which I spent the entire summer investigating, have prompted me to make significant changes that I have not yet fully clarified and for which I regrettably have little time at the moment.*"

In *Prometheus*, he summarized his controllability experiments with a single sentence, writing that his efforts to make the glider easier to handle and usable in a wider range of wind conditions included various mechanisms for adjusting the shape of the wing as desired, which likely included wing warping mechanisms. Just a few years later, a wing warping system was further developed by the Wright brothers until it was ready to be patented, a decisive step toward their three-axis-controlled aircraft. In August 1895, Lilienthal wrote to Wolfmüller:

> "*You are entirely correct. The shift in the center of gravity must be greater than a person can accomplish when gliding in the wind with large wings. As the simplest method of balancing the lifting capacity of the two wings, I recommend allowing the wings to rotate around the longitudinal axis. I have found this to be the safest method compared to any others. It is also the method that is used by birds.*"

Fig. 7 A similar flight position, but the angle of attack is already negative. Lilienthal is countering the position by shifting his weight strongly backward. Might he have engineered this flight situation deliberately? The photograph may have been taken shortly before the slat opens. Photo: Neuhauss. © Otto-Lilienthal-Museum. All Rights Reserved

But Lilienthal appears to have suspended his own experiments with wing warping once again. The fact that his hands were not free to operate a mechanical twisting mechanism no doubt played a decisive role. Instead, Lilienthal wanted the mechanisms to be controlled by a willow ring around his hips. This system worked to complement the effects of weight shifting, which Lilienthal continued to consider the key to controlling his flying machines. However, Lilienthal's flight position did not allow large hip movements. It would have been difficult to adjust the twisting levers at the lower endpoints of the cross frame sufficiently finely with the hips. Wolfmüller seemed to have been right: mechanisms should be operated with the hands, so a seated position within the aircraft was favorable.

For the tail that could be rotated right and left, we are forced to resort to guesswork, as Lilienthal never gave any further details. It was probably operated by pulling the tail sideways, connected by a cord running over the control struts. The tail likely returned to its default neutral position elastically

Fig. 8 Transporting the glider back to the take-off site, opposite to the direction of flight. Beylich is shouldering most of the load, with Lilienthal helping. Photo: Preobrazhensky, 1895. © Otto-Lilienthal-Museum. All Rights Reserved

when the cord was released. The pulling was presumably also performed with the hips.

All of these experiments, including tests with small control surfaces on the wings, were conducted with the *Experimental Monoplane*. This is documented by the photographs taken by Preobrazhensky in late July or early August 1895. The control surfaces were presumably shaped like a rounded trapezoid and turned to face into the wind. If necessary, they could be rotated by pulling a cord so that their broadside turned into the direction of flight to generate resistance and push back a surface that had advanced too far forward (Figs. 8 and 9).

Nothing is known about the various seating arrangements that Lilienthal mentioned. A prone position like the one chosen by the Wright brothers is not conceivable with Lilienthal's design, and explicitly stated as much. It would be desirable to achieve such a position at a later stage to reduce the air resistance, he said. But he needed his current flying position for takeoff, landing, and to allow himself to fall from the glider in dangerous situations.

Fig. 9 In the picture, the *Experimental Monoplane* is difficult to recognize. The photograph was taken by Preobrazhensky and published by Zhukovsky in 1897. The chimney of the steam brickworks and one of the workers' houses can again be seen in the background. © Otto-Lilienthal-Museum. All Rights Reserved

The Wrights followed up on this. They recognized that, even though Lilienthal had made thousands of flights, they only added up to a few hours of training across a period of five years. To solve the problem, they needed a way to stay airborne for longer. Wilbur Wright concluded his own assessment with the following:

> "*Although he experimented for six successive years 1891-1896 with gliding machines, he was using at the end the same inadequate method of control with which he started. His rate of progress during these years makes it doubtful whether he would have achieved full success in the near future if his life had been spared, but whatever his limitations may have been, he was without question the greatest of the precursors, and the world owes to him a great debt.*"

It appears that Lilienthal aborted his experiments with control mechanisms quite abruptly. A possible explanation for this can be found in the essay published in *Prometheus*, where a short passage on various control systems was followed by a curt dismissal:

> "*But I shall skip the successes achieved in this regard, as another principle yielded surprisingly favorable results.*"

This sounds more jubilant than resigned—Lilienthal had come across another idea that was a clear and probably unexpected success.

A Clever Idea: The Biplane

The idea that led Lilienthal to the biplane was as simple as it was ingenious. To transition from short glides to longer soaring flights, he needed to increase the lift of his flying machines considerably. Increasing the surface area of each wing, as he had attempted previously, required larger wingspans, making the aircraft more and more difficult to control in the air by shifting his weight. But if he stacked two small wings on top of one another, he could double the lift while keeping the same wingspan. The smaller the flying machine, the easier it was to control in strong winds.

In 1895, he made the following report in the *Journal for Airship Travel*:

> "*Before I started to build these double soaring apparatuses, I made small paper models according to this system to study the behavior of such a flying device during free motion in the air, and I used the results to build the apparatuses on a larger scale.*"

The very first attempts with small models already produced highly convincing results, achieving great stability in flight. Small-scale experiments showed that the upper wing was not disturbed if it was a sufficient distance from the lower wing. According to Lilienthal's findings, this distance needed to be around three quarters of the wings' chord (Fig. 1).

As the illustration shows, the models were equipped with three vertical partitions between the two wings. Lilienthal surmised that this would not only increase the model's strength, but also improve its stability in flight. Interestingly, the French brothers Gabriel and Charles Voisin and the

© The Author(s), under exclusive license to Springer Nature
Switzerland AG 2022
M. Raffel and B. Lukasch, *The Flying Man*, Springer Biographies,
https://doi.org/10.1007/978-3-030-95033-0_17

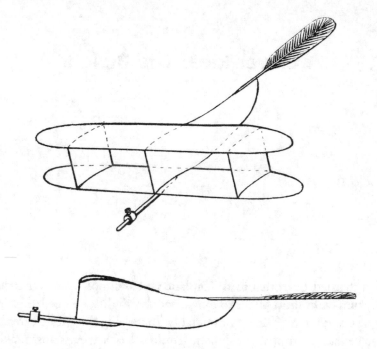

Fig. 1 Drawing by Lilienthal in the Journal for Airship Travel, 1895: Lilienthal's biplane model, which he wrote was characterized by excellent flight characteristics. When the wind was still, the flights were very consistent, always achieving the same flight duration when launched from the same altitude. The center of gravity could be regulated with an adjusting screw at the front of the fuselage spar. A bird's feather provided the tail

Brazilian Alberto Santos-Dumont would later use similar structures in their powered biplane designs.

The tail surface of Lilienthal's biplane model was a bird's feather. After adjusting the small sliding weight at the front of the model so that the center of gravity was just in front of the center of the wings, it achieved excellent flight performance. According to Lilienthal, the models danced around in the air as if they were alive:

> "Since they are so lightweight, their flight path is almost horizontal, or even increasing when the wind is weak. With a perseverance that is almost poignant, the little fliers attempt to direct themselves into the wind, sometimes hovering in the air at a certain point, following magnificent circles, as adeptly as a buzzard. After floating for an extended period, they finally land gently on the ground."

In the first so-called *Small Biplane*, Lilienthal's first man-carrying biplane, each wing had a surface area of 9 m². This allowed Lilienthal to achieve

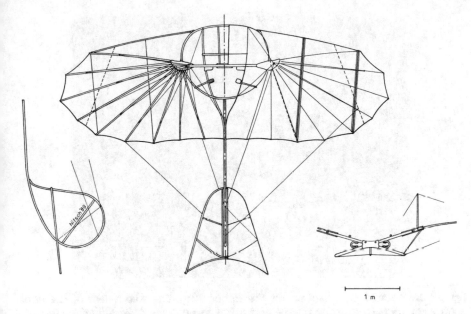

Fig. 2 Reconstruction drawing of the *Storm Wing Model*. © Otto-Lilienthal-Museum. All Rights Reserved

a total area of 18 m^2 with a wingspan of just five and a half meters. The *Sturmflügelapparat* from the previous year was used for the lower wing.

The way in which the upper surface was attached was very straightforward (Fig. 2). It was placed on two bamboo sticks and tied to the lower wing with string. The bamboo sticks were placed on the hinge pockets in the same holes that had previously held the struts for the monoplane's upper bracing. The upper wing was not foldable, consisting of two rigid halves splinted together in the middle of the wing.

The maximum depth of the upper wing was two meters (Fig. 3). There was only one profile rail on each half, but an additional arched willow rod was placed against the ribs from below. On each side of the two connecting bamboo sticks, there was a small tensioning block for the upper bracing. This detail was copied over from the lower wing, but Lilienthal removed it in the large biplane, as it probably proved unnecessary with the rigid upper wing.

But the first biplane, as well as the larger one that followed it, benefited from another decisive advantage that was not intentionally planned by Lilienthal: in Lilienthal's monoplanes, the pilot was close to the center of gravity of the glider. In the biplane, part of the lift was generated above the pilot's head. The center of gravity was significantly lower than the center of lift. This gave the biplane additional inherent stability in flight, provided that it was properly trimmed. Like a dandelion seed or a parachutist, the pilot was

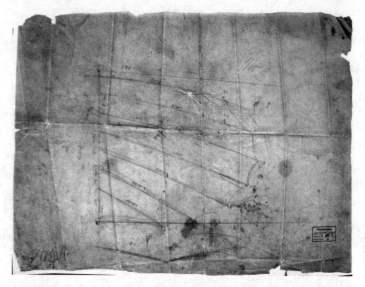

Fig. 3 Draft sketch by Lilienthal for the upper wing. The two halves of the upper wing were connected in the middle and installed onto the *Sturmflügelapparat* that had been built in 1894. The two wings are connected with bamboo struts and tension cords. The upper wing was not designed like a bat's wing and could not be folded, only separated into two sections. © Otto-Lilienthal-Museum. All Rights Reserved

now hanging below the wing supporting him, a geometric configuration that would be adopted by every modern hang glider.

The biplane was just as easy to handle as the monoplane, which encouraged Lilienthal to try it in winds of up to 10 m/s (meters per second).[1] According to him, this produced the most interesting results out of any of the flight attempts.

Even at a wind speed of 6–7 m/s, he was able to take off from the top of the hill almost horizontally, without a run-up. In stronger winds, he frequently even gained some altitude. At the apogee of these flight paths, the biplane sometimes came to a standstill for several seconds, making things a bit easier on his eager photographers:

> "On these occasions, I felt very clearly that if I laid slightly to one side, making a circle, and advancing with the rising air, I would remain lifted in the air. The wind itself wants to initiate this movement, and my main task in the air is to prevent any turning to the right or the left, because I know that the hill from which I took off is behind me and under me, and I would be experience a rough impact if I allowed myself to begin circling".

---

[1] 10 m/s are equivalent to 20 knots or 22 mph

Neuhauss and Fülleborn took several short-exposure photos of the flights with the small biplane on Monday, October 7, 1895. From meteorological records, we know that there were strong winds of 6–8 m per second from the west-northwest at noon, and winds of 4–6 m per second from the west-southwest in the evening on that particular day.

At that point, the two biplanes had become Lilienthal's favorite aircraft. But, in addition to the more complex structure, experience and practice was required to master the larger and heavier flying machines. Consequently, for customers and students, Lilienthal continued to only offer his monoplane, the *Normalsegelapparat.*

The name of the *Sturmflügelapparat*, on which the smaller biplane was based, was chosen by Lilienthal himself. It can be found on a draft sketch preserved in the *Deutsches Museum* in Munich, probably drawn by Lilienthal's technician, Paul Beylich: *Storm Wing Model*, April 24, 1894. 9 m^2 surface area. The somewhat more complex, offset shape of the frame ring, which had been removed from the *Normalsegelapparat,* suggests that the supporting frame ring was reused from an even earlier model. After 1894, the supporting structure of the gliders was more or less standardized. The cross frame with the frame ring and the "cockpit", the brackets for the pilot's arms, the hinge pockets for the movable wing ribs, and the tail unit socket were identical in every glider that was actually built. All later devices used this stable, inverted suspension-bridge inspired frame, to which the wings, tail units, and then the second wing were attached simply with flexible bracing. In Lilienthal's flying machines, the load-bearing bracing is below the frame and carries the pilot during flight. The upper bracing, which simply consists of hemp cord, stabilizes the wing shape when the glider is unloaded and on the ground.

The details of the wing shape also differed from other designs. The front edge of the wing is pulled far forward and reinforced with a thin willow rod. In the *Normalsegelapparat*, the front edge is just a tensioned rope. The inner profile rail is inserted up to the front edge. A seam on the rear edge of the wing shows that the glider was trimmed by extending it. Since the pilot's position is fixed, further trimming can only be accomplished by modifying the wing surfaces.

Interestingly, the *Sturmflügelapparat* is the only flying machine other than the *Normalsegelapparat* that has been preserved. But it is also the only glider without any contemporary photographs. Presumably, Lilienthal's plan to master stronger winds with the smaller wingspan was not met with much success. The lack of lift of the smaller wings may have meant that any flights simply weren't long enough to be photographed. A key reason why this particular glider was preserved is therefore that it was used in Lilienthal's biplane, which was surprisingly successful, and of which there are many photographs.

Fig. 4 Lilienthal with the 18 m² *Small Biplane* on October 7, 1895 at the foot of the *Fliegeberg*. The upper wing has a significantly smaller wingspan than the lower one, and the distance between the wings is about 1.5 m. A thick seam in the covering can be seen at the rear edge of the lower wing. © Otto-Lilienthal-Museum. All Rights Reserved

On August 4 or 5, Langley visited from Washington, D.C., and watched as Lilienthal, who had explicitly requested that nothing about it be discussed publicly at that point, demonstrated the biplane with flights on the *Fliegeberg*. Langley reported about the biplane to his assistant Augustus M. Herring in a letter from Berlin on August 6. According to his statements, the biplane flights achieved equivalent results to the monoplane that Lilienthal had also demonstrated that day.

The favorable flight results soon inspired Lilienthal to build a second, larger biplane. This aircraft was built according to the same principle, starting with a 6.7-m-wingspan *Normalsegelapparat* as the substructure, supplemented by a second upper wing, this time built larger (Figs. 4, 5, 6, 7, 8, 9, 10 and 11).

Lilienthal reported about this second, larger biplane in the *Journal of Airship Travel* in October 1895:

> "*I then made a larger, double apparatus with a total wing area of 25 square meters. It works very well in calm weather, but the 7-meter wingspan again makes it difficult to control in stronger winds.*"

Fig. 5 As can be seen in this photograph, the upper wing of the *Small Biplane* was only equipped with one profile rail on each side. Outside, however, Lilienthal also placed a curved willow rod against the ribs from below. Two small tensioning braces extending the bamboo struts between the wing surfaces are installed on the upper wing. © Otto-Lilienthal-Museum. All Rights Reserved

Neuhass and Fülleborn returned to take more photographs, documenting multiple flights of Lilienthal's new larger biplane on October 19, 1895. According to the records of the *Meteorological Institute* in Potsdam, the wind on that Saturday was very weak, which is consistent with Lilienthal's statements. At noon, a wind speed of just 1 m per second was measured from the west, and the wind in the evening came from the southwest at a speed of 1.5–3.5 m/s.

Lilienthal hoped that the large biplane would be able to replicate the soaring ability of birds and enable flights of much greater duration. He viewed circular flight as a symbolic achievement toward this goal. The idea of flying in a circle for the very first time seems to have exerted an almost magical appeal. In *Prometheus*, he wrote:

"Once I or another experimenter succeed in flying a full circle for the very first time, the event should be regarded as one of the most important achievements on the path to complete flight. Only then will we be able to fully exploit the living power of the wind."

Lilienthal's 15-m-high *Fliegeberg*, allowing him to take off from a height of around 12 m, was a temporary solution that allowed him to practice more often in the flat areas surrounding Berlin. To fly farther from greater heights, he transported the large biplane to Gollenberg the following year, returning to the airfield where he had achieved flights over a distance of 250 m back in 1893 and 1894. His goal was now to switch from his training slope, the *Fliegeberg*, back to the higher hills with his new gliders, where he hoped to be able to make better use of the natural wind from a greater altitude. Wood took several photographs of Lilienthal's flights at this location on August 2, 1896 and even tried out the glider for himself. Wood published a detailed account of his experience in the *Boston Evening Transcript* shortly after Lilienthal's death.

"*On the following Sunday I met him by appointment at the Lehrter station in Berlin. He was accompanied by his [seven]teen-year-old son, whom he always took with him, and a man servant to assist in putting the machine together. We steamed out of the city and across the flat green fields just as the sun was rising, and after a ride of a couple of hours alighted at Neustadt on the Dosse. Here we were met by a comfortable farm wagon and driven over the twenty-odd miles which lie between Rhinow and the railroad. A brisk wind was blowing, and the storks were sailing over the fields on each side of the road in search for food for their young on the chimney-tops. Them he watched with great interest, calling them his teachers, and drawing my attention to the various methods practiced by them in preserving their*

Fig. 6 Original print by A. Regis' studio in Berlin of a snapshot showing gliding flight with the *Small Biplane*. Thanks to Regis and the amateur photographers Dr. Neuhauss and Dr. Fülleborn, we have a whole series of photographs of flights with the *Small Biplane* from the autumn of 1895. © Otto-Lilienthal-Museum. All Rights Reserved

Fig. 7 Regis' studio also made an enlarged reproduction of this photograph by unknown means. © Otto-Lilienthal-Museum. All Rights Reserved

Fig. 8 Most contemporary publications of photographs of Lilienthal's flights were printed in the popular science magazine *Prometheus*. In this magazine, Lilienthal published many essays about his progress in aviation technology, second only to the *Journal of Airship Travel*, the official journal of the VFL. This photograph was printed in *Prometheus* in October 1895. © Otto-Lilienthal-Museum. All Rights Reserved

Fig. 9 Lilienthal added the following comment to this photograph of a flight published in *Prometheus*: "*Even at wind speeds of 6–7 m per second, the large wing area of 18 m² could carry me almost horizontally from the top of my hill without a run-up into the wind. When the wind was stronger, I simply allowed myself to be lifted off the hill and slowly glide into the wind. You can see the strength of the lateral motion that sometimes occurred...*"

equilibrium when flying and alighting, which I had never noticed before. We had a hurried lunch in the little inn at Rhinow, where his arrival always causes a hum of excitement among the peasants; the flying machine was brought out of the barn, and loaded on the wagon, and we drove away to the mountains, which are two or three miles from the village.

A more ideal spot for flying could hardly be conceived. Rising abruptly from the level fields is a long range of high, rounded hills, varying from one to three hundred feet in height and thickly carpeted with grass and deep, spongy moss. The slopes vary in steepness and face all possible points of the compass, so that one can always find a suitable declivity facing the wind.

The machine was laid out on the grass and put together. It was one of the new models, consisting of two large lateral wings measuring twenty feet from tip to tip, and an upper wing or aeroplane. The material was thin, strong cotton cloth, tightly stretched on a frame of bamboo. A rectangular wooden frame which fitted around the body a little above the waist supported these wings and the duplex tail, consisting of a horizontal and vertical wing joined together on the end of a curved bamboo pole in the rear of the machine. The upper aeroplane was supported some six feet above by means of two vertical rods of bamboo, and firmly fixed by a great number of tightly stretched strings or guys. So perfectly was the machine fitted together that it was impossible to find a single loose cord or brace, and the cloth was everywhere under such tension that the whole machine rang like a drum when rapped with the knuckles. As it lay on the grass in the bright sunshine with its twenty-four square yards of snow-white cloth spread before you, you felt as if the flying age was really commencing. Here was a flying machine, not constructed by a crank, to be seen at a country fair at ten cents a head, or to furnish material for encyclopedia articles on aerial navigation, but by an engineer of ability; and embodying the results of eight years of successful experimenting. A machine not made to look at but to fly with.

We carried it to the top of the hill, and Lilienthal took his place in the frame, lifting the machine from the ground. He was dressed in a flannel shirt and knicker-bockers, the knees of which were thickly padded to lessen the shock in case of a too rapid descent, for in such an emergency he had learned to drop instantly to his knees after striking with his feet, thus dividing the collision with the earth into two sections and preventing injury or strain to the machine.

I took my place considerably below him by my camera, and waited anxiously for the start: he faced the wind and stood like an athlete waiting for the starting pistol. Presently the breeze freshened a little; he took three rapid steps forward and was instantly lifted from the ground, sailing off nearly horizontally from the summit. He went over my head at a terrific pace, at an elevation of about fifty feet, the wind playing wild tunes on the tense cordage of the machine, and was past me before I had time to train the camera on him. Suddenly he swerved to the left, somewhat obliquely to the wind, and then came what may have been a forerunner of the disaster of the next Sunday. It happened so quickly and I was so excited, at the moment that I did not quite grasp exactly what happened, but the apparatus tipped sideways as if a sudden gust had got under the left wing. For a moment I could see the top of the aeroplane, and then with a powerful throw of his legs he brought the machine once more on an even keel, and sailed away below me across the fields at the bottom, kicking at the tops of the haycocks as he passed over them. When within a foot of the ground he threw his legs forward, and notwithstanding its great velocity the machine stopped instantly, its front turning up allowing the wind to strike under the wings, and he dropped lightly to the earth. I ran after him and found him quite breathless from excitement and the exertion. He said: 'Did you see that? I thought for a moment it was all up with me. I tipped so, then so, and I threw out my legs thus and righted it. I have learned something new; I learn something new each time.'

Though I had read many articles about Lilienthal, and had seen numberless photographs of him in the air, I had formed no idea of the perfection to which he had brought his invention, or of the precision with which he managed it. I have seen high dives and parachute jumps from balloons, and many other feats of skill and daring, but I have never witnessed anything that strung the nerves to such a pitch of excitement or awakened such a feeling of enthusiasm and admiration as the wild fearless rush of Otto Lilienthal through the air. The spectacle of a man supported on huge white wings, moving high above you at racehorse speed, combined with the weird hum of the wind through the cords of the machine, produces an impression that can never be forgotten.

A few moments' rest were necessary before carrying the machine once more to the hilltop, and we sat on the grass and discussed the incidents of the first flight. The grasshoppers clicked about on the cloth wings, and Lilienthal laughed at them, and said that they loved to jump about on the smooth white surface, that they were his only passengers and he frequently heard them hopping about on his machine when he was in the air. The wind had freshened a trifle and a shower was seen coming across the plain. We crawled under the wings, together with a swarm of peasant children who had flocked from the neighboring farms to watch "Die Weise Fledermaus" [the white bat], and kept quite dry during the cloud-burst. The sun came out presently, and by the time we had reached the top of the hill the wings were quite dry. Once more he took his place in the frame and sailed away, the children running screaming after him down the steep hillside and falling over each other in their excitement. Of this flight and the subsequent ones, I was fortunate enough to secure some excellent pictures, the last ones that were ever taken of the man.

Toward the end of the afternoon, after witnessing perhaps half a score of flights, and observing carefully how he preserved his equilibrium, I managed to screw up courage enough to try the machine. We carried it a dozen yards or so up the hillside, and I stepped into the frame and lifted the apparatus from the ground. The first feeling is one of utter helplessness. The machine weighs about forty pounds, and the enormous surface spread to the wind, combined with the leverage of ten-foot wings, makes it quite difficult to hold. It rocks and tips from side to side with every puff of air, and you have to exert your entire strength to keep it level. Lilienthal cautioned me especially against letting the apparatus dive forward and downward, which is caused by the wind's striking the upper surface of the wings and is the commonest disaster which the novice meets with. The tendency is checked by throwing the legs forward, as in alighting, which brings the machine up into the wind and checks its forward motion. As you stand in the frame your elbows are at your side, the forearms are horizontal, and your hands grasp one of the horizontal cross-braces. The weight of the machine rests in the angle of the elbow joints. In the air, when you are supported by the wings, your weight is carried on the vertical upper arms and by pads which come under the shoulders, the legs and lower part of the body swinging free below.

I stood still facing the wind for a few moments, to accustom myself to feeling of the machine, and then Lilienthal gave the word to advance. I ran slowly against the wind, the weight of the machine lightening with each step, and presently felt the lifting force. The next instant my feet were off the ground; I was sliding down the aerial incline a foot or two from the ground. The apparatus tipped from side to side a good deal, but I managed to land safely, much to my satisfaction, and immediately determined to order a machine for myself and learn to fly. The feeling is most delightful and wholly indescribable. The body being supported from above, with no weights or strain on the legs, the feeling is as if gravitation has been annihilated, although the truth of the matter is that it hangs from the machine in a rather awkward and wearying position. My second attempt was not so successful, the wind getting under the left-hand wing and tipping the machine until the tip of the other wing dragged on the ground. No damage was done, however, and I felt quite satisfied with my first attempt.

On the way back, Herr Lilienthal talked about his experiments and his plans for the future. Certain features of his machine he has patented, though his experiments have been made without any money-making view. The machines cost about 500 marks or $125 to build, not much more than a first-class bicycle, and when made in quantity can be made very much cheaper. He told me that he hoped to sell his American patents and asked me if I thought he could get $4000 for them. His plan was to build, in or near Berlin, a sort of flying rink, with an artificial slope which could be turned so as to always face the wind. Here people could come and hire machines and learn to use them, commencing with small elevations and gradually going higher up the slope, as practice gave them skill. He hoped to get people, particularly athletes, interested in the sport, for with a wide interest would come improvements. The present bicycle is not the work of a single man, but the result of years of experiment and thought given by many men. It must be the same with the flying machine. If the unfortunate death of the pioneer does not deter others from experimenting along these lines, and it does not seem to me that it should, the results accumulated by him will not be lost and he will not have given up his life in a vain cause. He has made thousands of flights in safety and felt absolute confidence in his ability to control his machine in any ordinary wind, and his accident was merely one of those liable to come to anyone engaging in any of the popular outdoor sports.

Undoubtedly the danger element is greater in this sport than in most others, but with improved apparatus it can be made, in my opinion, as safe as tight-rope walking, which is really not so very perilous when you know how. Lilienthal is certainly the first man of modern times who has navigated the air for any distance without the aid of a balloon. Maxim's wonderful airship, so far as I know, has not yet been run in free flight, though developing astonishing speed and buoyant power on its track. The former practiced soaring flight against the wind. Without any movement of the wings, the latter drives his aeroplane through the air by an engine and screws. Lilienthal had the advantage of being part of his machine, as it were, feeling every change of plane and instinctively correcting it with a motion of the body. He thus slowly acquired the skill necessary to keep the apparatus level with varying wind pressure.

The aeroplane at present is an unstable machine, requiring the agency of a human mind to keep it in equilibrium, and the necessary skill can be best acquired with small machines fastened directly to the body. I do not think that anyone who has experienced the difficulties encountered in keeping one of these small machines in equilibrium would venture to carry Mr. Maxim's aeroplane into the air, the balancing being effected by rudders put into action by opening or closing throttle

Fig. 10 Lilienthal with the second, *Large Biplane*, with a total wing area of 25 m². A *Normalsegelapparat* with an area of 13 m² served as the lower wing section. Support rods for the two bamboo struts connecting to the upper wing are attached to the cross frame, running diagonally upward. The upper wing section had an area of 12 m². Lilienthal had removed the small struts on the upper wing of the *Small Biplane*, as they were unnecessary. Photo: Neuhauss 1895. © Otto-Lilienthal-Museum. All Rights Reserved

valves. It would be, to my mind, like trying to ride a steam bicycle fifty feet high, perched in a small cab on top of the huge wheel, with a row of valves for driving and steering, without having had any experience with a small machine. The small machine is undoubtedly the one to commence with, and what Lilienthal's machine needs to increase the safety factor is some means of loosening things when struck by a sudden gust. I use this expression in a very broad sense, of course, and possibly the desired effect could be obtained by a ballasting device, but flying with the machine as it now is like trying to sail a boat with the mainsheet fast. It can be done, but it is risky in squally weather."

Fig. 11 Lilienthal is gliding above the photographer in the *Large Biplane*. Here, we can again see that the upper wing is only connected in the middle. For transport, it was separated into two halves. The cross frame, the hinge pockets, and the frame circle form the heart of Lilienthal's flying machines. The wing outline, ribs, profile rails, and seams in the covering are clearly identifiable, as well as the cross frame, frame circle, and tail. The wingspan of the upper wing is only slightly smaller than that of the lower wing. Photo: Neuhauss, October 19, 1895. © Otto-Lilienthal-Museum. All Rights Reserved

There are three known photographs taken by Wood on August 2, 1896, very similar to the photographs of the *Fliegeberg* from 1895. These photographs are the last ever taken of Otto Lilienthal. They are also the only photographs of biplane flights at any training location other than the *Fliegeberg* in Lichterfelde. Finally, they are the only photographs at all of Lilienthal's flights on the Gollenberg. The restaurant at the nearby *Herms Inn* in Stölln where Lilienthal regularly stayed was later renamed *Zum Ersten Flieger* (*To the First Flyer*) in his honor (Figs. 10, 11, 12, 13, 14, 15, 16 and 17).

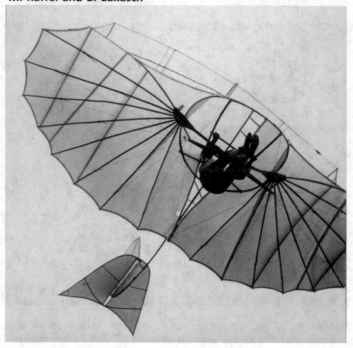

Fig. 12 Similar photograph position, also taken on October 19, 1895, showing the flying machine against the sky, like an X-ray. Lilienthal says that he was able to arrange the most suitable position for the photograph with the photographer while in the air. Here, the *Large Biplane* is photographed from almost directly underneath, and these elements are clearly visible. The rear strut carrying the tail unit is also visible, extending to the rear edge of the frame ring. At its end, there is a pin inserted into a hole. The tail unit is secured by a pin at the rear edge of the wing. An interesting detail: there is a small opening in the covering that allows the pin to be inserted on the ground without having to lift the device. © Otto-Lilienthal-Museum. All Rights Reserved

Fig. 13 Gliding flight with the biplane in a western direction on October 19, 1895, as seen from the *Fliegeberg*. This photograph was also widely circulated during Lilienthal's lifetime. The flight appears safe and elegant, reminiscent of modern hang-gliders. Photo: Neuhauss, October 19, 1895. © Otto-Lilienthal-Museum. All Rights Reserved

Fig. 14 The Austrian Wilhelm Kress, who visited the *Fliegeberg* in 1896, felt that the upper wing of Lilienthal's biplane was attached primitively and carelessly. Kress gave a critical but accurate description of the design: *"This upper, very primitively built wing consisted of two halves connected in the middle and held together by a few simple pins. Two rods protruded upwards from the lower wing, which was itself properly and robustly built. The upper wing was then attached to these two rods by inserting their tips into two rather wobbly metal cones on the upper wing. Finally, the two wings were connected and stiffened simply with cords."* Photo: Neuhauss, October 19, 1895. © Otto-Lilienthal-Museum. All Rights Reserved

Fig. 15 Lilienthal with the *Large Biplane* at the top of the *Fliegeberg*. This picture shows that two men were needed to carry the aircraft up the hill. Paul Beylich probably also assisted with take-off by supporting the wing until Lilienthal jumped. Even under weak wind conditions, the large glider with its wingspan of almost 7 m could not be kept in balance by Lilienthal alone at the top of the hill. Photo: Krajewsky. © Otto-Lilienthal-Museum. All Rights Reserved

Fig. 16 Landing maneuvers with the biplane were no different than with the mono-plane. In the final phase of the flight, Lilienthal leans his upper body far back, lifting the front of the device. This drastically increases the angle of attack, breaking off the air flow, and decreasing the speed of the aircraft. He then touches down with his feet stretched out and makes a few quick steps. For the sake of safety, his knees are protected with thick pads in case he hits the ground too hard and does not manage to land on his feet. Photo: Neuhauss, October 19, 1895. © Otto-Lilienthal-Museum. All Rights Reserved

Stölln

Der Gollenberg

Fig. 17 Historical photograph of the Gollenberg near Stölln. Today, it is covered with trees. © Otto-Lilienthal-Museum. All Rights Reserved

Unfinished Ideas and August 9, 1896

In 1896, every aspect of Lilienthal's life seemed to pick up even more momentum. News of his experiments reached many different countries. Could his dream of flying as a sport really come true? His articles were printed in French, Russian, and English. Many countries were talking about the flying Prussian (Fig. 1).

It was at this point that Lilienthal began to receive new orders for his *Normalsegelapparat* from remote locations such as Washington and Moscow. Was his vision of a sport uniting the world's nations in peaceful competition prophetic after all? In his last lectures, he talked of ideas beyond aviation itself, emphatically restating his predictions about the elimination of borders and wars.

> "*What cultural progress we would achieve if the open atmosphere could be used for general traffic, with no mountains, forests, waters, or swamp to hinder our travels. As you might already have realized, notions such as borders between countries would completely lose their meaning, as it is impossible to cordon a country off up to the heavens. It seems difficult to imagine that tariffs and wars could still exist. The tremendous prosperity from international travel between the peoples of the world would ultimately mix all of the world's languages into one.*"

Supplementary Information The online version contains supplementary material available at (https://doi.org/10.1007/978-3-030-95033-0_18). The videos can be accessed individually by clicking the DOI link in the accompanying figure caption or by scanning this link with the SN More Media App.

M. Raffel and B. Lukasch, *The Flying Man*, Springer Biographies, https://doi.org/10.1007/978-3-030-95033-0_18

LA NAVIGAZIONE AEREA - L'UOMO CHE VOLA

Fig. 1 Title page of *La Tribuna* (Italy), illustrated sunday insert, october 14, 1894, full-page colour engraving based on a photograph taken by Anschütz in 1893, subtitled *La Navigazione Aerea—L'uoma che vola* (*Aviation—The Flying Man*). © Otto-Lilienthal-Museum. All rights reserved

These words can be found in a lecture manuscript from 1895, delivered to an unknown audience. In June 1896, he wrote the following in the *Neue Hamburger Zeitung*:

"My efforts to turn these practical flight exercises into a stimulating sport have failed thus far because there were no suitable training grounds near Berlin. For this, we require an exposed hill with a flat conical shape and a height of 25 to 30 meters, allowing flights of 250 to 300 meters in all directions. I imagine that, once our gymnastically inclined youths, in addition to rowing and cycling, begin launching themselves from the heights of hills to build their strength, flying at the speed of an arrow to reach a distant goal in competition with one another, a sport of gliding will develop and solve once and for all the problem of human flight."

Lilienthal's counterparts in America had understood this dream. Lilienthal was proud to have been invited to open a flight school in America and saddened to have been forced to decline the opportunity, but he could not simply abandon his factory. A flight school with flight students from the USA was now being planned at his location in Berlin the following year. Lilienthal needed to find a patron like Means, except in Germany. Means had even asked about Lilienthal's specific wishes for building another *Fliegeberg*.

"Both the Russian government in Moscow and the private sector in Boston are currently examining the construction of a station for large-scale private flight tests. It would be a shame if a similar endeavor were not undertaken in our own country due to a lack of entrepreneurship."

But his efforts in Germany had not yet been crowned with success.

"Of course, if there was already money to be made from flying, people would undoubtedly cease to be so indifferent to its mysteries. But speculation, that great driving force underlying technical progress, has not yet found its footing here."

However, the great trade exhibition planned to celebrate the 25th anniversary of Berlin as the nation's capital would offer an opportunity to advocate for aviation. The exhibition was originally supposed to be a world's fair. After Germany's rival France had caused a sensation in 1889 with their *Paris Exposition Universelle* and the opening of the *Eiffel Tower*, Berlin was anxious to prove themselves equal as Europe's leading industrial metropolis. At the Kaiser's request, it was renamed the *Berlin Industrial Exposition* instead of a world's fair. With an area of 900,000 square meters, it was larger than previous world's fairs. From May 1 to October 15, 1896, a total of 3780 exhibitors presented the latest developments in science, industry, business, and sport in Treptow Park, as well as scenery depicting an entire city, called *Old Berlin*. Seven million visitors were ferried between the highlights of the exhibition along an electric tramway.

As was disturbingly common at world's fairs of the era, the exposition featured a *Colonial Exhibition*, an ethnological exhibit that would be extremely disconcerting to modern visitors. It featured a so-called *Negro Village* of 100 inhabitants from Africa, shown off and stared at over a period of seven months. A far more positive byproduct of the event was the *Archenhold Observatory*, built around the world's longest moving refracting telescope and which has been preserved to this day. According to Eduard Spranger, later a professor of philosophy and education at the *University of Berlin*, the exhibition was responsible for elevating Berlin from a provincial city of Europe to a global metropolis.

Otto Lilienthal also presented his company at the exhibition. In the main industrial building, he exhibited steam engines, safe steam boilers, and wrought-iron transmission discs. A steamboat was operated on the nearby Spree river, powered by a coiled tube boiler made by Lilienthal's. At the landing stage of the *Kaiserschiff*, a foghorn with a three-part chord siren built by Lilienthal was installed and sounded every hour. Lilienthal had submitted a design to the organizers of the exhibition for a balloon hall that could be used to host aeronautical demonstrations. The central rotunda for the balloons was designed with take-off tracks for gliding flights on three sides. But the proposed design was never implemented. It is not known whether this design was ever seriously considered for the exhibition. Regardless, the flight demonstrations that he had originally planned were not approved, so he had to content himself with demonstrations and lectures on the ground. Flying machines were not presented at the trade exhibition, but tethered balloons were released, and the ascent of a dirigible by Hermann Wölfert was announced.

On July 10, 1896, Lilienthal gave a lecture on *Practical Flight Experiments*: The *Berliner Zeitung* wrote (Fig. 2):

"A very interesting lecture was given by the aforementioned on Wednesday in the chemistry building of the exhibition in front of a packed room, presenting his method and the results achieved so far. The quintessence of his observations and experiments culminates in the conclusion that the question of flight can only be brought to a final solution gradually, step by step. Unfortunately, there is too much publication and too little experimentation in this field. If our young people were as enthusiastic about the sport of flying as they are now about cycling, the growing competition would soon inspire flying machines to be perfected, just as we have seen happen with bicycles in recent years. [...] These interesting and instructive flight

Der Hörsaal des Chemie-Gebäudes. Vortrag des Herrn Otto Lilienthal über „Praktische Flugversuche".

Fig. 2 Illustration for the lecture by Lilienthal on June 10, 1896 published in a volume about the exhibition in 1897

experiments have been successfully photographed. The photographs, enlarged and projected to the room, aroused lively interest from the visibly animated audience, as did a showing of the large flying apparatus."

The year 1896 was also full of significant events for Lilienthal in other areas. On May 13, his play *Moderne Raubritter* premièred at the *National Theater*. But, at first, he did not want to be credited as an author, and chose a pseudonym, Carl Pohle, based on his mother's maiden name. Lilienthal's real name was only printed in the program after his death (Fig. 3).

Since Lilienthal had helped transform the *Ostend Theater* into the *National Theater*, most of the plays staged there had been classical pieces. The drama

Fig. 3 A part-time theater director: swearing an oath to found a *Volksbühne* (*people's theater*). Otto Lilienthal in the middle, (co-)director Max Samst to his left, and actor Richard Oeser (with the top hat), 1892. © Otto-Lilienthal-Museum. All rights reserved

written by Otto himself was an exception: a social critique and period play set in Berlin. The subtitles announced *Stories from the Industrial Life of Berlin in Five Acts (Eight Pictures) Based on Real Events Adapted to the Stage.*

Progress was also under way in aviation. Lilienthal's wanted to do more than simply teach others to fly—he wanted to continue his own progress as a pilot and aircraft designer. On the back of a letter dated June 8, 1896, he sketched a new airplane with thick wings covered on both sides. The letter was from someone named A. Merck in Berlin, offering information about a water velocipede that the sender claimed to have invented. Lilienthal was no doubt interested in this idea because his factory was also working on a pedal-operated boat. Designing an aircraft with thick, dual-surface wings marked a departure from his focus on collapsible gliders with wings covered on just one side. But it is not entirely surprising, as Lilienthal had already wondered in the *Journal for Airship Travel* in 1895 whether it might be a mistake to focus only on bat-type wings when building flying machines simply because they are easier to execute in practice. And he had of course already remarked in his book on bird flight that it was unlikely to be a coincidence that the wings of birds were considerably thicker at the leading edge due to the presence of "arm" bones. His own measurements had concluded that thick wings are not an obstacle to flight; on the contrary, they can provide more lift and considerably improve performance at slow speeds (Fig. 4).

Two decades later, in 1910, Hugo Junkers filed a patent for a thick-profile cantilever wing in Germany, a precursor to his historic series of metal airliners.

We have little more than sketches and guesswork to tell us about Lilienthal's final ideas and his final, incomplete flying machine project. It is

Fig. 4 Lilienthal, studying the dynamic soaring flight of storks in Vehlin (Brandenburg) in the company of the photographer Dr Fülleborn. An account of the journey was given in his article *Our Teachers for Hovering Flight* (1895). © Otto-Lilienthal-Museum.

probably the device that remained in the factory, which Beylich said was burned under the factory's steam boiler after Lilienthal's death. In sketches, Lilienthal had designed a new wing profile that was considerably thicker in the middle section of the wing. That thicker center section was formed by two spars and eight ribs installed parallel the direction of flight. This was connected to an external section of the wing derived from previous designs: a small collapsible bat wing, covered on just one side, with radial ribs and a hinge pocket at the end of the two wing spars. Lilienthal wanted the middle section, which was covered on both sides, to also be foldable. To make this possible, joints were added in the middle of the spars so that they could be folded together, like the upper and lower parts of a bird's wing.

In another sketch by Schauer shortly before Lilienthal's death, the elevator unit is controlled by a cable that runs to a lever behind Lilienthal's back.

"It is worth noting that, when Lilienthal died, in addition to the second, half-finished wing flapping aircraft mentioned above, there was another aircraft with considerably thicker wings covered on both sides that was almost ready for testing."

Schauer stated this on the record, albeit more than three decades after Lilienthal's death. While it is consistent with Beylich's statements, Schauer and Beylich were in close contact on many questions in the period that followed. We only know for certain that Lilienthal had not finished his intensive development work on the aircraft. Given the fact that it was flying that led to his untimely death, it is understandable that Lilienthal's aviation legacy wasn't immediately recognized. His loss was a severe blow to both his family and his employees, and, in their anguish, they arranged for his workshop to be destroyed. There was never any doubt that Gustav would take care of Otto's family and business, and this had likely been arranged in Otto's will. With Agnes Lilienthal's four underage children, the day-to-day business of the factory, the industrial exhibition, and everything else, there was neither the time nor the inclination to worry about Lilienthal's place in aviation history.

In 1896, Otto's flying experiments had generated such great interest from the general public that the Kaiser himself wanted to see the flights in person and announced a visit. On the occasion of such visits, it was customary for the Kaiser to inquire whether the host had any special request that the Kaiser might grant. Schauer discussed this with Otto, suggesting that he should formally request government funding for his research, as was happening in other countries, especially given how much he'd already spent out of his own pocket. Otto was happy to learn of the Kaiser's interest, but didn't want to ask for money for his flying research—rather, he was more interested in getting government support for his theater as an educational institution. But the Kaiser's visit was not to be.

After Wood's visit on August 2, 1896, Lilienthal invited him on short notice to join him again the following weekend at Stölln. Lilienthal presumably intended to allow the American to fly the standard monoplane before deciding whether or not to purchase one. This explains why, on August 9, 1896, Lilienthal, rather than advancing his experiments with newer designs, was flying the tried and tested monoplane that he had flown thousands of times before. As fate would have it, Wood had another commitment and was unable to make it to Stölln. Believing that his guest had simply been delayed, Lilienthal allowed the Hamburg train to depart before taking the next train from Magdeburg to Rathenow. This was not really a detour, as it followed practically the same route as the train from Hamburg to Neustadt on the Dosse.

Lilienthal made his first flight of the day on the Gollenberg before noon. Beylich had travelled out earlier on Saturday to make some modifications to the *Normalsegelapparat*. The wind was blowing from the east at a speed of 3 m per second, and Lilienthal's first flight was successful, extending far out onto the plain below the Gollenberg. For the next flight, Lilienthal asked Beylich to measure the time with a stopwatch. With an almost horizontal flight path, it looked like he was going to achieve a long flight after his jump. In his essay *About my Flight Experiments*, Otto had talked several times about how his glider would occasionally come to a standstill, hovering for a few seconds whenever the wind suddenly picked up and matched the minimum speed of his flying machine. Even in calm weather, he would encounter thermals, sudden updrafts that would lift and decelerate his glider. As long as he had sufficient altitude, Lilienthal could maintain control by shifting his weight in response to a gust or a thermal. But he wanted to do more than simply recover from these phenomena—he wanted to take advantage of them, learning to soar. Lilienthal was flying an altitude of about 15 m when he was suddenly caught in a particularly turbulent thermal. The glider shot upward, and rapidly decelerated, almost to a halt. Lilienthal threw his legs and upper body forward to force the glider to pitch down and gain airspeed. It's likely that, at this point, the wind subsided as he'd come out of the thermal, which meant that Lilienthal had unwittingly over-corrected. The glider dove almost vertically to the ground, crashing onto its right wing. Beylich ran to the crash site, and found Lilienthal lying unconscious underneath the glider. Beylich later described the incident as follows:

"Suddenly, the bird stopped in the air – a sudden calm in the wind, perhaps – Lilienthal swung his feet a little to continue moving forwards, but the apparatus

suddenly flipped over and plummeted into the ground headfirst. I ran over and pulled the man out from the wreckage. 'Lilienthal!', I called, 'What's wrong!' But he gave no answer – he was unconscious. I quickly grabbed my bottle of seltzer and rubbed his forehead. Then I sent two of the boys standing around to find a doctor and a cart. Suddenly, Lilienthal opened his eyes. 'What was the matter, Beylich?', he said to me. 'Crashed! Well, it's not that bad,' Lilienthal reassured me. 'It happens sometimes. Let me rest for a while, then we'll try again.' 'We will do no such thing,' I shouted. 'The machine is completely broken.' He simply nodded. He was not in pain, but he could not move a limb. […] Then the doctor came. He examined him in the field and stated: 'Nothing is broken, everything is intact.' - 'Well,' I thought to myself, 'that would truly be a miracle in a crash.' So, we took the patient to the Herms hotel in Stölln, where he lost consciousness again. I immediately went to Berlin to inform his brother. After midnight, I arrived at Lehrter station and walked two and a half hours to Lichterfelde. I raised Gustav Lilienthal from his bed. Gustav arranged for transport to Berlin the next day, August 10, 1896. Without removing him from the camp bed on which he had been laid the day before, he was brought to Neustadt in a horse-drawn carriage, then transported to Berlin in a freight wagon."

The local doctor quickly concluded that Lilienthal's condition was severe, with paralysis clearly indicating a spinal injury. Ernst von Bergmann was the leading medical expert of the day, so the choice of hospitals in Berlin was clear. During the trip, with the doctor and his brother by his side, Lilienthal became drowsy and lost consciousness. On arrival at the Bergmann hospital, the forerunner to today's *Charité Berlin University Clinic*, doctors could to nothing more than confirm the diagnosis of the doctor from Rhinow—Lilienthal was beyond help.

Otto Lilienthal died that same day, Monday, August 10, 1896, at about half past four in the afternoon. He was just 48 years old. Doctors at the Bergmann hospital performed an autopsy, which revealed a fracture of the third cervical vertebra. Lilienthal's crash had broken his neck with full force.

Four days later, on August 14, the world's first aviator was buried in the old Lichterfelde cemetery on Lange Street. The crashed monoplane was transported to Berlin for a police investigation. The damage to the glider appeared relatively minor, especially given the severity of Lilienthal's injury. Only the right wing had broken when it hit the ground. The two back cushions that Lilienthal added after his accident in 1894 to prevent his body from tilting backward too far are clearly visible inside the cross frame (Fig. 5).

The police investigation was concluded, finding that no third party was at fault. According to the testimony of Anna and Gustav Lilienthal, the crashed glider was burned along with other devices in the factory. The upper wing of the biplane used in Stölln was also not preserved.

Fig. 5 Crashed glider with visible damage to the right wing. The flying machine appears to have been transported back to Berlin and incompletely set up in the yard of Otto Lilienthal's factory. *Photo*: Regis, August 11, 1896. (▶ https://doi.org/10.1007/000-6yk) © Otto-Lilienthal-Museum. All rights reserved

When Lilienthal was buried, hundreds of people joined the funeral procession behind his coffin: relatives, friends, neighbors, leading representatives of the VFI., and workers from his factory. Moritz von Egidy, chairman of the *Egidy Association* of which Lilienthal had been a member, placed a wreath, as did representatives of the German *Free Land Association*, the *Free Land Settlement Cooperative*, and members of the old *Ostend Theater*. All of the mourners praised Lilienthal for his dedication to social issues and his cultural involvement.

Egidy, a former military officer turned influential social ethicist who espoused the concept of a modern, less dogmatic form of Christianity, was a source of inspiration for the Lilienthals, as was Theodor Hertzka and his views on free land, free trade, and a free economy. Lilienthal had exchanged correspondence with Egidy on how aviation could contribute to peace and understanding among different peoples of the world. Egidy wrote what is perhaps Otto Lilienthal's most beautiful obituary:

"The engineer Otto Lilienthal had a hand in every serious cultural endeavor of modern times. He was both a clear thinker and a man of action; with a tender heart. Anyone who knew him will hold the memory of this excellent man dear."

Otto's sudden death had serious ramifications for his family, and the shadow of his loss still lingered decades later. Agnes was of course the most devastated. Only two months after the death of her husband, she lost her father too, also in an accident. Otto's family remained at the house in Boothstrasse 17 for a few years longer, but their financial situation gradually became more and more hopeless. The mortgages that Otto had taken out on the house could no longer be paid. The family was forced to give up the house.

Despite every obstacle and difficulty, Agnes successfully prepared her children for a working career. The eldest, Otto, was an excellent student at the high school in Lichterfelde. Later in life, however, he suffered from mental health issues and died in 1916 at the *Bodelschwingh Institute*, where Gustav worked as a master builder. His younger brother Fritz began studies at the *Technical University of Berlin*, just like his father before him. Unlike his father, however, who studied at a time where technology received little recognition as a subject, Fritz received the scholarship that Otto had been refused. The girls Anna and Helene became teachers and later lived with their families in Lichterfelde.

In 1910, Gustav republished a second edition of Otto's book with his own additions. The first edition had been published before Lilienthal began his actual test flights. Since then, a lot had been added to Otto's story that was well worth telling. In the meantime, the Wright brothers had shown the world that powered flight was possible, including a demonstration in Berlin in 1909. Interest in Otto's book had grown, and the second edition sold significantly better than the first. A third edition, edited by the aerodynamicist Ludwig Prandtl, who left out Gustav's additions, was published in 1939.

Orville and Wilbur Wright always acknowledged that their success was a continuation of Lilienthal's work. On April 20, 1911, a lawyer, Dr. Heinrich Adams, inquired about the possibility of meeting with Agnes Lilienthal on behalf of Wilbur Wright:

"I would like to take this opportunity to suggest in confidence that Mr. Wright also intends to offer his service to the family of the revered old master of the art of flying."

In a letter dated December 2, 1911, Orville Wright wrote:

"Dear Madam: As you already know we have great admiration for the work of your deceased husband the late Otto Lilienthal, and it has been a matter of much regret to us that we never had the pleasure of his personal acquaintance. He was a great man. As a token of our appreciation of him we beg that you will accept

the enclosed exchange for one thousand dollars with our best wishes for a Merry Christmas and a Happy New Year.
   Yours truly Wright Bros."

The letter contained a check for $1000, the equivalent of more than $25,000 in 2021. Agnes replied with her thanks on Christmas Eve (Figs. 6 and 7):

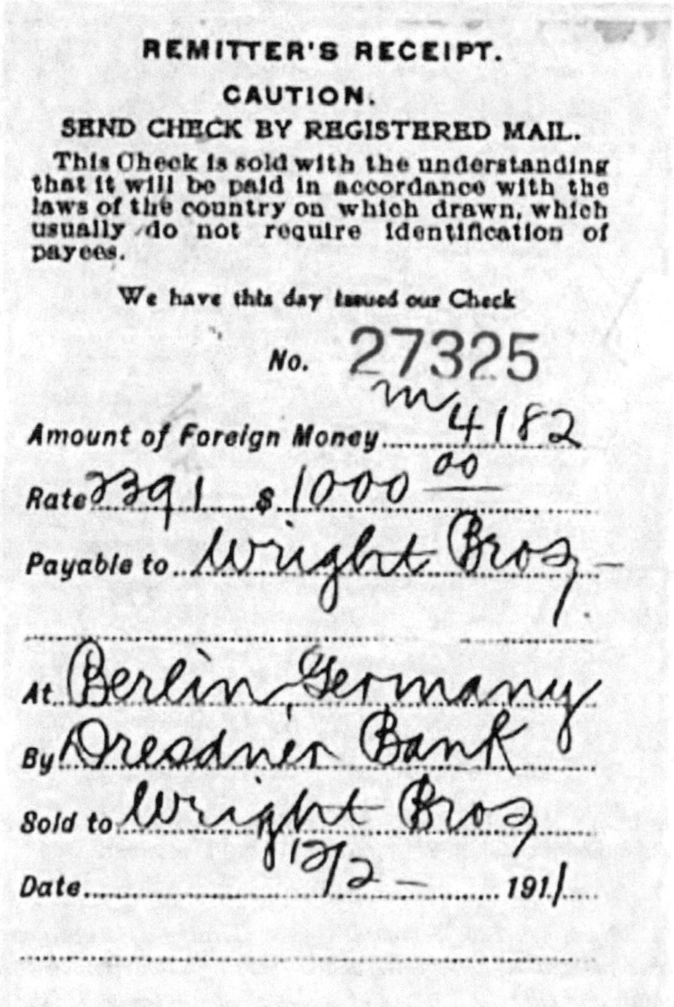

Fig. 6 Check for $1000 to Agnes Lilienthal, *Library of Congress*, manuscript division, Washington D.C. box43 wright. © Library of congress, manuscript division, Washington. All rights reserved

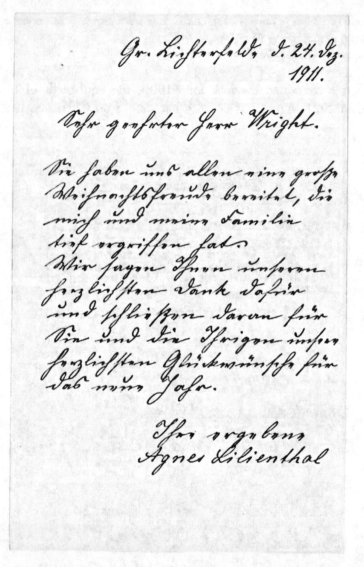

Fig. 7 Letter from Agnes Lilienthal to Orville Wright, *Library of Congress.* © Library of congress, manuscript division, Washington. All rights reserved

"*Dear Mr. Wright, you have bestowed a great Christmas joy upon us that has deeply moved me and my family. We would like to give our heartfelt thanks for your generosity and offer the warmest of wishes to you and yours for the new year. Yours sincerely, Agnes Lilienthal*"

In 1910, the city council of Anklam invited Otto Lilienthal's siblings and their families, Gustav from Berlin and Marie from New Zealand, to return

In diesem Hause wurde am
23. Mai 1848 der Ingenieur
Karl Wilhelm Otto
Lilienthal,
der Altmeister der
Flugtechnik, geboren.

Fig. 8 Memorial plaque at the house where Otto Lilienthal was born, unveiled in 1910. © Otto-Lilienthal-Museum.

and attend the unveiling of a memorial plaque at their birthplace. This was the first ever official tribute to Lilienthal. While the house was destroyed in World War II, the plaque was recovered from the rubble and is now displayed in the town museum (Figs. 8 and 9).

In 1914, a privately-funded memorial was erected in Lichterfelde, which is now a district of Berlin, by the renowned architect and sculptor Peter Breuer. The memorial is in a park halfway between Lilienthal's house and the *Fliege-berg*, and was later described as "*the most beautiful ever built for an engineer.*" A plaque on the memorial quotes Leonardo da Vinci (Fig. 10):

"*From the crest of the hill the great bird will take his first flight, filling the universe with wonder, all chronicles with his fame. And eternal glory be to the place he was born.*"

In 1920, the area around the *Fliegeberg* was turned into a park, and a Lilienthal memorial was erected at its foot. The *Fliegeberg* itself was then converted into a Lilienthal memorial based on plans by Fritz Freymüller, the city planning officer of Steglitz, in 1932. Freymüller added other simple geometric shapes to the conical mound of sand: a ringed roof, columns, cuboids, and a sphere. There are seventy-five steps leading up to the top of the *Fliegeberg*, with four stone benches inviting visitors to make a contemplative stop while at the top (Figs. 11 and 12).

Otto's brother Gustav Lilienthal died one year after the memorial was completed, just two days after the *National Socialists* seized power in Germany, and efforts by the country's new leaders to use Lilienthal's name and achievements for propaganda purposes began.

Fig. 9 The monument for Otto Lilienthal made of polyester resin, inaugurated in 1982 in Anklam, combines Lilienthal's glider with modern glider technology. In addition to the life data, it bears the quote from his book as an inscription: *"Just apply the power of thought—also you will be borne by the ether."* © Otto-Lilienthal-Museum. All rights reserved

Fig. 10 The Lilienthal memorial erected in 1914 in Bäkepark in Lichterfelde. *Photo* Eggert © Otto-Lilienthal-Museum. All rights reserved

In 1936, as part of the ongoing propaganda campaign, all of Germany's research institutions were combined into the newly founded *Lilienthal Society for Aviation Research*. Otto Lilienthal's grave was transformed and inscribed with his alleged final words, "*Sacrifices must be made*" (Fig. 13).

In 1936, Hermann Göring, the Commander in Chief of the German Luftwaffe, unveiled an award for outstanding scientific achievements in aviation—surprisingly, for foreign scientists. The award was known as the Lilienthal ring, a men's ring with a large agate gemstone set in gold and

Fig. 11 Inauguration of the Lilienthal memorial on the *Fliegeberg*, august 10, 1932.

Fig. 12 The Lilienthal park today. *Photo* Wittig © Otto-Lilienthal-Museum.

Fig. 13 Otto Lilienthal's grave after national socialist redesign 1940/today honor grave of the city of Berlin. *Photo* Rausch © Otto-Lilienthal-Museum.

Fig. 14 The *Lilienthal Ring*, only ever awarded once, a unique piece, can now be found in the *Otto-Lilienthal-Museum*. © Otto-Lilienthal-Museum.

engraved with a portrait of Lilienthal by the renowned glass artist Wilhelm von Eiff (Fig. 14).

This award was presented for the first and only time in 1938 to Sir Roy Fedden, the President of the *Royal Aeronautical Society* of Great Britain and the Director of the *Bristol Motor Works*, the leading manufacturer of radial aircraft engines. World War II began soon afterward. Fedden appears to have been courted over several visits of Germany and, interestingly, was given access to military and secret projects, which he reported on back in England.

In the years after the war, the public's view of Otto Lilienthal shifted in both sides of Germany. He was honored for his humanistic and peaceful ideals. Many sports associations were named after him, and today many streets and schools in cities have followed suit. The German *Society for Aeronautics and Astronautics* (DGLR) also added the names of the two German aerospace pioneers, Lilienthal and Oberth, to its official title.

Between 1965 and 1975, a new airfield was built in the western part of Berlin, on a site that had originally been used as an airfield for the Berlin Airlift. The airfield's modern architecture would attract international recognition for its Hamburg-based architects, Gerkan, Marg, and partners. As

Fig. 15 The intriguing architecture of the Berlin-Tegel *Otto Lilienthal* airport, composed of hexagons. *Photo* Eggert © Otto-Lilienthal-Museum. All rights reserved

city development grew up around it, the Berlin-Tegel *Otto Lilienthal* airport closed operations in November 2020 when a new airport was opened outside of Berlin (Fig. 15).

To Fly is Everything—Lilienthal 125 years Later

In 2016, the *German Aerospace Center* (DLR) conducted wind tunnel measurements of a museum reproduction of Lilienthal's *Normalsegelapparat* on the occasion of the 125th anniversary of Lilienthal's first flights. After Lilienthal's death, most people believed that his aircraft simply could not be flown stably and safely, and that it had been Lilienthal's extreme fitness and skill that had allowed him to fly it at all. Even though he had flown his glider thousands of times before the final crash, it was this failed attempt that defined many people's opinion on the matter.

After evaluating all the documents available in the *Otto-Lilienthal-Museum* in his hometown of Anklam and examining the original gliders preserved in the world, a flyable replica was built and tested. It was the first comprehensive and scientific investigation of the aerodynamic properties and flight mechanics of the world's first successful heavier-than-air aircraft. The tests showed that the glider was stable in both pitch and roll with good flying characteristics and a glide ratio of about 4:1 (Figs. 1, 2 and 3).

After DLR's successful wind tunnel tests, which proofed the gliders flight stability, the test pilot reported:

Supplementary Information The online version contains supplementary material available at (https://doi.org/10.1007/978-3-030-95033-0_19). The videos can be accessed individually by clicking the DOI link in the accompanying figure caption or by scanning this link with the SN More Media App.

M. Raffel and B. Lukasch, *The Flying Man*, Springer Biographies, https://doi.org/10.1007/978-3-030-95033-0_19

Fig. 1 Aerodynamic quality of the 20-kg lightweight glider with a wingspan of 6.7 m was put to the test in one of the world's largest and most modern wind tunnels, the DNW-LLF in Emmeloord, the Netherlands. *Photo* German aerospace center (DLR). (▶ https://doi.org/10.1007/000-6yr)

Fig. 2 Markus Raffel preparing tethered flight tests in the largest European wind tunnel. DNW-LLF. *Photo* Jan Vetter

Fig. 3 Learning experience during first attempts to control the device in artificial wind. *Photo* Jan Vetter (▶ https://doi.org/10.1007/000-6yn)

"*I was obsessed by the idea of flying Otto Lilienthal's wonderfully designed monoplane after the successful wind tunnel tests of DLR. These tests proved the stability of Otto Lilienthal's foldable monoplane, his Normalsegelapparat, using a scientific approach. However, the question remained, can an inexperienced pilot of average fitness foot-launch this craft, fly it safely and perform coordinated flare landings?*"

Bringing the glider to flight in 2018 was extremely hard work. After successfully testing it with sand ballast, he started tethered flights on a 16 × 16-foot platform. The glider was attached to a trailer towed by a car. This allowed us to gain some experience with the glider without too much risk of bodily harm. In a second step, we performed flights at a limited altitude of two meters while pulled by a rope winch. The test pilot flew at speeds of up to 30 mph over soft grass. Due to his self-imposed altitude limit he had to maintain the winch tension until shortly before landing. Just as Lilienthal's 50-foot cone shaped hill enabled him to always launch directly into the wind, our winch was also set up so that every flight was made into the wind. During these winch flights we struggled, but finally succeeded in finding the right trim and learned to control the glider reliably in roll (Figs. 4 and 5).

Fig. 4 Lilienthal training simulator. *Photo* Jan Vetter.
(▶ https://doi.org/10.1007/000-6yp)

Fig. 5 More demanding training of lateral control during winch flight. *Photo* Arno Trümper. (▶ https://doi.org/10.1007/000-6yq)

Fig. 6 Realized: free flights at Marina Beach close to Monterey. *Photo* Jan Raffel.
(▶ https://doi.org/10.1007/000-6ym)

After these successful tests with the monoplane glider tethered on a rolling platform and the short winch flights, the first free flights took place in California in 2018. In order to achieve this safely, we teamed up with Andy Beem, owner of *Windsports*, a long-running hang-gliding school in Los Angeles. He was the one who recommended flying at Dockweiler Beach and in the Monterey area: a dune with constant wind, not too crowded, with the right slope, and in a country where flying uncertified ultralight aircraft with weight shift control is legal. The location couldn't be any better (Fig. 6).

Excited by these flights of Lilienthal's patented monoplane glider after more than one hundred years, we requested an authentic replica of Otto Lilienthal's *Grosser Doppeldecker* (*Large Biplane*) from the Otto-Lilienthal-Museum. Both Raffel and Beem flew this authentic biplane glider on the California Coast near Monterey in July 2019 (Fig. 7).

Team members reported about the challenges:

"In April 2019 we shipped [a replica of] another of Otto Lilienthal's masterpieces, the Large Biplane, from Germany to the NASA Ames Research Center in Mountain View, CA. Earlier, with months of effort, the glider had been carefully built by the skilled experts of the Otto-Lilienthal-Museum. With help from a handful of hardy volunteers, we moved it out of the box, carried all of the gear over the dune, and set the glider up on top of the same dune from which Markus flew the monoplane glider a year earlier.
This year we had just three days: everything had to be exactly right from the very beginning. The wind was steady, slowly increasing from 10 mph (16 kph) before

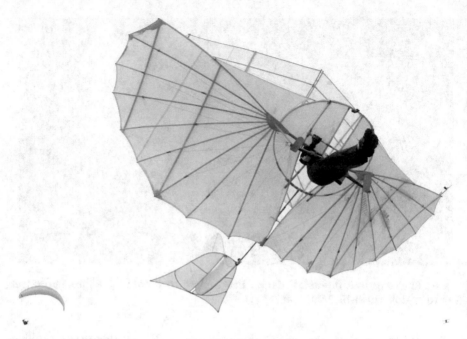

Fig. 7 Andrew Beem flying the Lilienthal biplane. *Photo* J. T. Heineck.
(▶ https://doi.org/10.1007/000-6ys)

noon to a maximum 14 mph in the afternoon. At first, we ventured only a few yards up the dune, and Markus, with his monoplane glider experience, made the first attempt. He started very carefully and slowly, but the glider flew beautifully right from the beginning. It glided steadily and pitched up as soon as Markus shifted his weight a few inches backward. On the first day, the flights lasted just three or four seconds, but we were all excited and satisfied: one could easily see that the glider flew very safely and steadily and reacted surprisingly well to every weight shift of the pilot. Both Andy and Markus flew that day.

The next day started out even better. We began early after a night in Monterey. Everyone knew what they were about. But something was wrong. All of us noticed that today the glider was flying left turns in spite of the pilot trying to compensate the inexplicable drift by shifting his weight to the right. Try as we might, we couldn't figure out the reason. Instead of bothering too much about this imperfection, all three of us enjoyed flying the glider from further up the dune. The left turn sometimes even helped to catch the crosswind that increased at the foot of the dune. George Reeves did his guest flight that day and flew and landed the glider right on his first try. We discovered later that one of the strings keeping the two upright bamboo posts in position had shifted and led to a misalignment. These two posts separate the upper from the lower wing. They were no longer parallel due to the misalignment and the upper deck was pointing a few degrees to the left.

Our third day of flying the biplane was set for one week later. It was the end of July and Markus had had time to set the glider and find and correct the askew string attachment at the post. The result was exhilarating: the glider flew perfectly straight. It was still wonderfully sensitive to pitch control and didn't do steep turns on lateral control but reacted immediately and predictably. It became clear why Lilienthal, the flying man, had known that he had found the answer to a key question: how to combine a large wing surface and keep the controllability of a monoplane. He had struggled before whenever he tried to increase the wing surface by increasing the span width. Then, in October 1895, he had finally overcome this problem: the "Large Bi-Plane" flew majestically at low winds. For the very first time, he handed the glider over to a man who had no flight experience, to give it a try. The American physicist Robert W. Wood met Otto Lilienthal at the Gollenberg and flew and landed the glider on his own. He immediately made plans to buy one of the gliders and later described his afternoon with Lilienthal and his flight experience in an article. The two had met one week before Lilienthal's fatal crash. One could say that Andy's flights were similar to the flight of this American flight enthusiast, 124 years ago. It took Andy only one or two attempts to fly higher, longer and more controlled than Markus, who had spared no effort to bring the glider to this point: authentically made, perfectly trimmed, and transported to just the right place to be tested. Andy weighed 55 pounds less and had 25 years more experience in hang gliding. And it showed: Andy was able to perform perfect flights. Not high enough for full turns but always safe, always stable, and under full control, with flights lasting up to 14 seconds and sailing more than one hundred yards."

In October, Raffel and Andy Beem visited the *Kitty Hawk Kites* hang gliding school at Nags Head, at the Outer Banks, North Carolina and made tethered flights in a replica of the 1902 Wright Glider built for the *2003 Centennial of Flight* by *The Wright Experience*, a group of experts in aeronautical engineering and vintage aircraft restoration. The idea of simultaneous flights of the Lilienthal and Wright gliders was hatched and well supported by John Harris, the owner of *Kitty Hawk Kites* and his team of flight instructors lead by Billy Vaughn.

Now there was nothing standing in the way of the idea of simultaneous flights of the machines of Wright and Lilienthal. The team members reported:

"On the 14th of December 2019 we carried the authentic replicas of the 1902 Wright glider and Lilienthal's biplane out to the dunes of Jockey's Ridge State Park in North Carolina and assembled both craft. As we moved them to the top of the dune, the grey clouds blew away to reveal a perfect blue sky. The wind was blowing from the south-west which forced us to launch on a dune with only little slope. This didn't allow starting the glider without being tethered. However, with one string that replaced some of the downhill force and two strings required to secure the large span of the Wright glider in the vicinity of the ground, it was supposed to work."

The pilots Andy Beem (Lilienthal) and Billy Vaughn (Wright) took their places and, after waiting for favorable winds, they launched. In that moment two seminal aircraft, both groundbreaking critical steps to the invention of the airplane, were in the air, together. It was obvious that each one represented a step in the overall development: Lilienthal working directly from bird anatomy, the Wrights coming after, their design pointing to the airplane.

"I've flown the Wright glider a number of times, but I am very pleased to have had the presence of mind to look out the left wing and through the structure of that aircraft and see the Lilienthal glider flying next to me. It dawned on me after the fact that this was a view that no one ever had. It is very humbling. The Lilienthal Glider was much more intuitive. I am a hang glider pilot, and I was right at home. It felt fantastic and simple, and it landed just like I thought it would. It was great!"

said Billy Vaughn, pilot at *Kitty Hawk Kites*.

"You can spend a lot of time reading, you can go to the museums, and you can see the aircraft, but there is something that is just not replaceable about actually flying the aircraft: to be in a tactile way connected to the folks who invented them. That is another humbling experience and I think it is a valuable one. It is one you can't get in any other way."

Andy Beem said after the flights (Figs. 8 and 9):

Fig. 8 Rendezvous at a historic location: Wright (1902) and Lilienthal (1895) on the outer banks, 2019. *Photo* Ignite films. (▶ https://doi.org/10.1007/000-6yt)

Fig. 9 7 s for eternity: a documentary on the DLR's *Normalsegelapparat Project* (26 min). *Photo* Filmhaus Berlin. (▶ https://doi.org/10.1007/000-6yv)

"*It was one of the best days of my life. I got to fly an authentic replica of an Otto Lilienthal Glider and I got to fly side-by-side with the 1902 Wright Glider, being part of history of showing Otto Lilienthal's influence on the Wright Brothers. I am very impressed with Otto Lilienthal's engineering and design of his glider using weight shift for take-off, control and landing and also seeing how the Wrights solved some of the flaws in the design of Otto Lilienthal Glider and made their Glider more controllable and ultimately developed the world's first airplane.*"

Appendix 1: List of Lilienthal's Known Flying Machines

Lilienthal's manned aircraft were experimental tools for him. They were changed and altered during the course of his research and testing. We know of nine different glider models from photos, while ideas and construction plans exist for other designs. Only the *Sturmflügelapparat* and restoration or fragments of four examples of the *Normalsegelapparat* have been preserved to this day.

1 Earlier Airplane Models

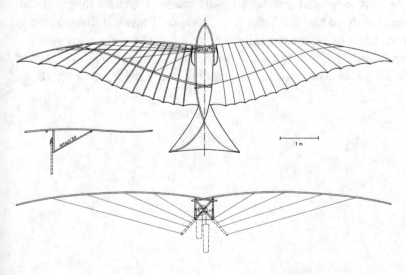

M. Raffel and B. Lukasch, *The Flying Man*, Springer Biographies,
https://doi.org/10.1007/978-3-030-95033-0_20

Lilienthal's first aircraft are not documented by photographs. Concept-studies and various model drawings exist. Presumably several flying machines of differing sizes were built. They were used to conduct various wind experiments and jumps (from a ramp and from a standing position).

Wingspan: 14 to 36 ft., wing area: 28 to 107 sq. ft.

2   Derwitz Glider 1891

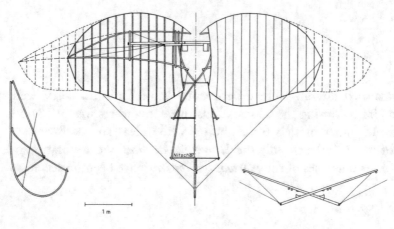

This is the first successful heavier-than-air manned aircraft in the world, flying distances of up to about 80 feet near Derwitz/Krielow in Brandenburg. During this series of experiments, Lilienthal reduced the glider's wingspan.

Wingspan: about 23 ft., wing area: about 86 sq. ft., wing curvature: 1/10 of length, max. length of wing: 5.6 ft., length of glider: 12.8 ft., weight: 18 kg.

3   Südende Glider 1892

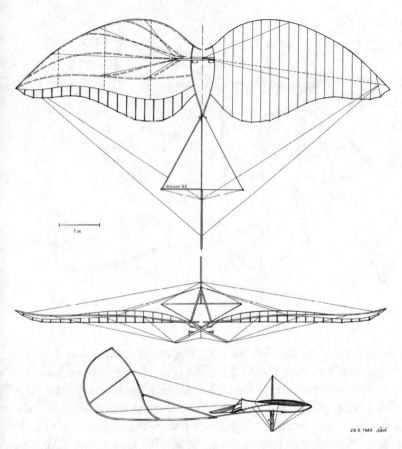

Lilienthal described it as a "gliding apparatus built over a framework". It is an aerodynamically sophisticated construction with fabric covering both sides of the wings. It flew up to a distance of about 90 ft (from a launching height of about 30 ft). From Lilienthal's papers we know the probable existence of another glider model using a similar construction but in a different size.

Wingspan: 31 ft., wing area: 158 sq. ft., wing curvature 1/20 of length, max. wing length: 8.2 ft., length of glider: 3.3 ft., weight: 24 kg.

4   Maihöhe-Rhinow Glider 1893

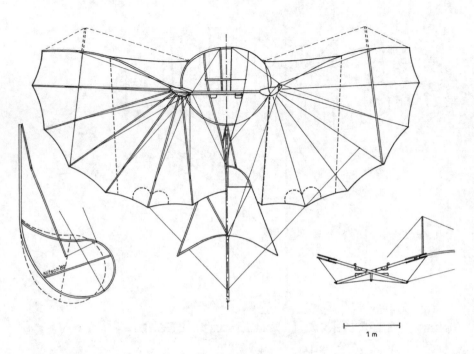

Referred by Lilienthal as the "convertible flight apparatus with 14 sqm [150 sq. ft.] wing surface" and *Modell 93*. This is the first glider of the new, convertible bat-like construction. Collapsed, the glider measures 6.6 by 10.5 by 1.6 feet. The wing profile could be changed by inserting different ribs or *Profile Rails*. This construction is legally protected with a patent. It is the basis of the later Normalsegelapparat. Near Stölln/Rhinow (Brandenburg), Lilienthal made flights covering 800 ft from a 200-foot-high hill.

Wingspan: 22 or 23 ft., wing area: 150 sq. ft., max. length of wing: 8.2 ft., length of glider: 14.3 ft., weight: 20 kg.

5   Small Ornithopter (Wing Flapping Apparatus) 1893–96

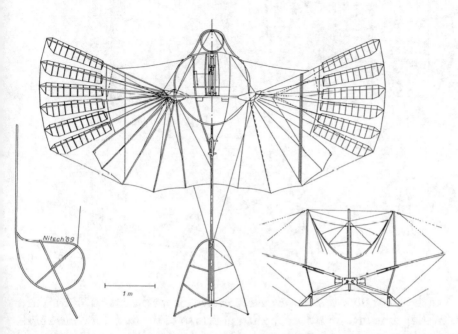

Building on his reliable glider design, Lilienthal attempted to add wing-flapping for motion power. Both muscle-power propulsion and motor-power propulsion were planned. In 1894 the first carbonic acid engine was ready for use. The results from the tests of the wing-flapping machine were not encouraging at first. Nevertheless, Lilienthal continued with his attempts to imitate the wing flapping of birds.

Wingspan: 22 ft., wing area: 129 sq. ft., max. length of wing: 8.2 ft., weight of engine: 5.5 kg, (approximately 10 kg including CO_2-cylinder).

6   Normalsegelapparat from 1894 Onward

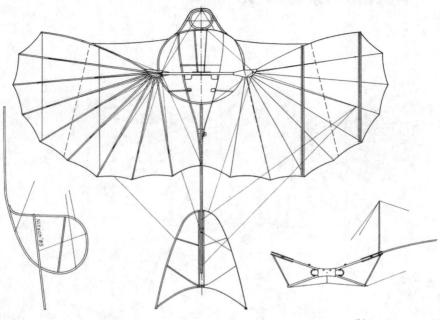

The design of the *Normalsegelapparat* evolved from the *Model Stölln*. There are existing drawings for two of the first gliders sold: the draft for *Seiler's glider* and for the *Modell Lambert*. Later, the *Normalsegelapparat* was refined into a biplane.

Wingspan: 22 to 23 ft., wing area: 140 to 146 sq. ft., max length of wing: 7.9/8.2 ft., length of glider: 16.1 to 17.4 ft., weight: about 20 kg.

7   Sturmflügelapparat Model 1894

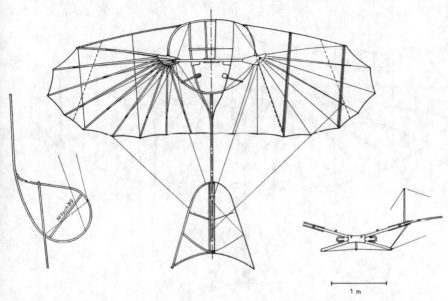

This glider was built with the same principles of construction as the *Normalsegelapparat*. The wings are reduced in size in order to withstand stronger wind. The glider was later refined into a *Small Biplane*. The original can be seen in the technical museum in Vienna.

Wingspan: 20 ft., wing area: 104 sq. ft., length of wing: 6.6 ft., length of glider: 14.8 ft.

8   Experimental Monoplane 1895

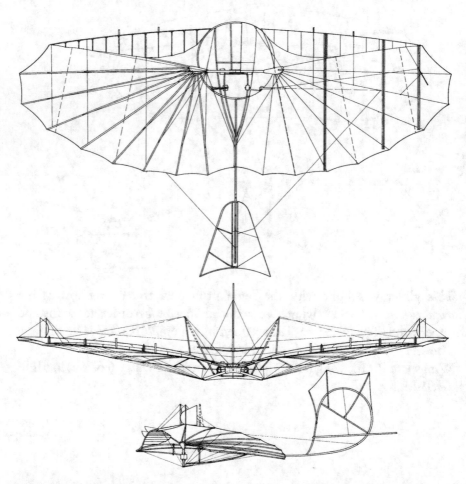

Lilienthal called this large monoplane an *Experimental Apparatus*. Different control mechanisms were tested on this monoplane glider. The conspicuous leading edge controller was intended to prevent rapid dives in the case of a negative angle of attack. A wing warping control and turn-able rear rudder and wing tip rudders (spoilerons) were also tested.

Wingspan: 29 ft., wing area: 224 sq. ft., max. length of wing: 9.8 ft., length of glider: 18.4 ft.

9   Small Biplane 1895

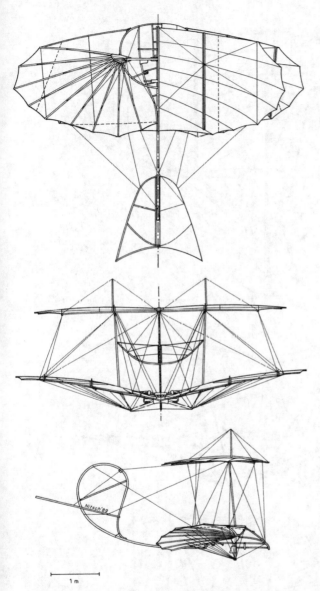

1 m

The *Small Biplane* was based on the *Sturmflügelapparat*. Lilienthal described his intention by saying: "*The biplane design has the same lifting capacity of a single wing with twice the span, but the shorter span is more responsive to changes in the center of gravity*". The results were convincing. The original lower wing is preserved in Vienna. Wingspan: 19.7/17.1 ft., wing area: 104/105 sq. ft., max. length of wing: 7.2/6.9 ft., length of glider: 15.7 ft.

10   Large Biplane 1895

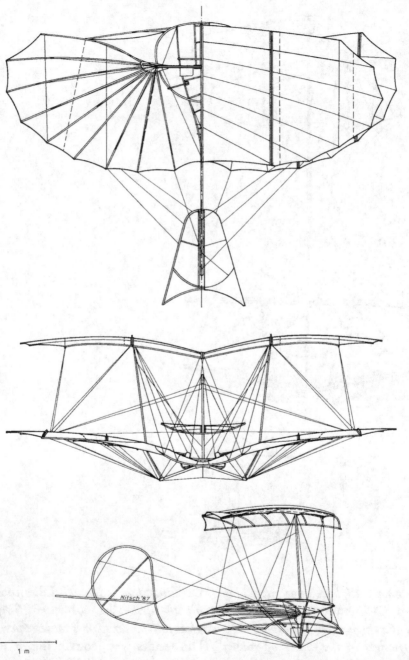

1 m

Feb 1987

The *Large Biplane* build from a *Normalsegelapparat* completed with a second wing after of the outstanding flying performance and controllability of the *Small Biplane*.

Wingspan: 21.6/20.7 ft., wing area: 146/112 sq. ft., max. length of wing: 7.5/7.5 ft., length of glider: 16.1 ft.

11   Large Ornithopter (Wing Flapping Apparatus) 1896

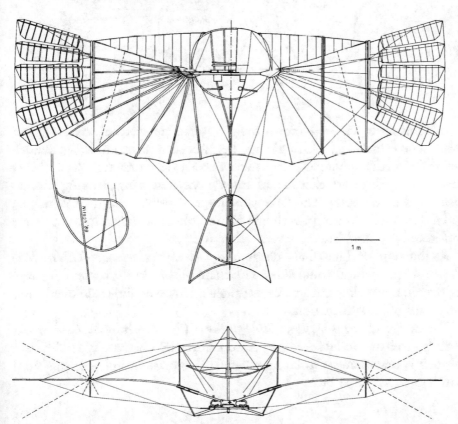

The *Large Ornithopter* was based on the wing-flapping machine of 1893, and was later outfitted with an engine. We assume the powered glider was completed but not tested. It had been shown in 1906 and 1907 in exhibitions, but it is not preserved today.

Wingspan: 27.9 ft., wing area: 188 sq. ft., max. length of wing: 8.2 ft., length of glider: 17.4 ft.

12   Other Designs and Constructions

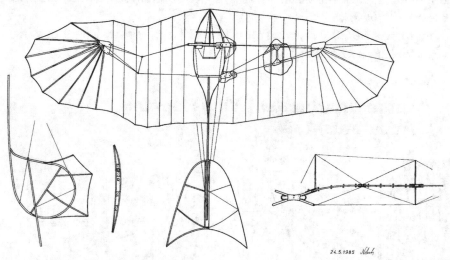

24.5.1985

In addition to the well-known designs, there are sketches, concepts and other source materials which cannot be associated with known gliders. A design for a large monoplane that survives today as a drawing was probably never built. There are sketches of muscle-powered wing flapping mechanisms and a helicopter. The designs of the collapsible wing (bat-principle) and a system of two wings, with one behind the other (tandem wings) were considered and used later by several other aviators.

At the time of Lilienthal's death, the *Gelenkflügelapparat* (*Glider With Jointed Wings*), which would allow a mechanical shift in the center of gravity for the pilot, was almost ready for test flights. However, little is known about the details of its construction.

The design of the *Kippflügel-Schlagapparat* (*Tilt Stroke Wing Glider*) was initially estimated to be from the period before 1893 because of its different form. It is now thought to be from 1896. Today we assume that this third ornithopter was never built.

Appendix 2: Otto Lilienthal—A Timeline

May 23, 1848.

Born in Anklam, Pomerania Province, German kingdom of Prussia.

1856–1864.

Grammar school in Anklam; His subjects included the study of birds, as well as mathematics under noted astronomer Gustav Spörer.

1864–1866.

Regional Technical School in Potsdam.

1866–1867.

Practical training at the *Schwartzkopff Company* in Berlin (mechanical engineering).

1867–1870.

Royal Technical Academy in Berlin.

from 1867.

First experiments, the results of which can be found in his book (published 1889) about the physical basics of human flight.

1870–1871.

Voluntary one year military service as a fusilier in the Franco-Prussian War during the siege of Paris.

1871.

Employment with the *Webers Company* (mechanical engineering firm), Berlin.

from 1872.

Construction engineer at the *C. Hoppe Machine Factory*, Berlin.

© The Author(s), under exclusive license to Springer Nature
Switzerland AG 2022
M. Raffel and B. Lukasch, *The Flying Man*, Springer Biographies,
https://doi.org/10.1007/978-3-030-95033-0_21

1873.

The Lilienthal brothers became members of the *Aeronautical Society of Great Britain,* Lilienthal gives his first public lecture about the theory of the flight of birds.

from 1874.

Systematic experiments on the force of air on artificial wings with models and kites and on the characteristics of natural wind.

1877.

Patent on a machine used in mining—the first of 25 patents by Lilienthal (among these were four aviation patents).

June 11, 1878.

Marriage to Agnes Fischer, daughter of a miner.

1879.

Birth of their son Otto (the first of four children).

Invention of what was later known as the *Anker-Steinbaukasten* (stone building blocks for children) together with Gustav Lilienthal.

1883.

Founded his own mechanical engineering company for boilers and steam engines in Berlin.

1886.

Membership in the *Deutscher Verein zur Förderung der Luftschiffahrt (German Society for the Promotion of Aeronautics)* in Berlin.

1889.

Original German publication of the book *Birdflight as the Basis of Aviation* (English translation published 1911).

1890.

Lilienthal introduces a 25% profit sharing scheme for the workers in his company, and conducts his first experiments with manned flying machines.

1891.

Jumps and first flights of a distance of about 80 ft in Derwitz/Krilow near Potsdam.

1892.

Improved flights with a new glider. Active interest in the *Berlin Volkstheater* (people's theater).

1893.

Construction of a flight station near his home. Commencement of his practice flights in the Rhinower Hills (Stölln/Rhinow near Neustadt/Dosse). Gliding up to a distance of about 800 ft.

Construction of several flying machines, among these a flapping wing machine driven by a motor.

Construction of his *Fliegeberg* which still exists in Lichterfelde, Berlin.
1894.
Serial production of the *Normalsegelapparat*.
1895.
Visits by aviation pioneers from different countries, among these S. P. Langley, secretary of the *Smithsonian Institution in Washington* and N. Y. Zhukovsky, aerodynamics expert from Moscow.
1896.
New experiments in the area of wing stroke.
August 9, 1896.
Crash after an unsuccessful attempt to fly a *Normalsegelapparat* through a turbulent thermal.
August 10, 1896.
Death in Berlin.

Appendix 3: Flight Testing Stability and Controllability of Otto Lilienthal's Monoplane Design from 1893

Markus Raffel ⓘ, **Felix Wienke, and Andreas Dillmann**

Nomenclature

A_f	Flight altitude ($A_f = 0 \triangleq$ to straight legs; feet on ground)
a_0	Lift slope in lift polar diagram
α	Aircraft's geometric angle of attack ($\alpha = 0 \triangleq$ vertical main frame)
α_0	Aircraft's absolute angle of attack ($\alpha_0 = 0 \triangleq$ zero lift)
α_e	Aircraft's trim angle of attack
C_D	Drag coefficient of aircraft
C_L	Lift coefficient of aircraft
$C_{M,ac}$	Longitudinal moment coefficient about aerodynamic center
$C_{M,cg}$	Longitudinal moment coefficient about center of gravity
D	Drag of aircraft
η	Geometric tail plane angle of attack
t	Time
Q	Pitch rate
M_{ac}	Longitudinal moment about aerodynamic center
M_{cg}	Longitudinal moment about center of gravity

Supplementary Information The online version contains supplementary material available at (https://doi.org/10.1007/978-3-030-95033-0_22). The videos can be accessed individually by clicking the DOI link in the accompanying figure caption or by scanning this link with the SN More Media App.

© The Author(s), under exclusive license to Springer Nature Switzerland AG 2022
M. Raffel and B. Lukasch, *The Flying Man*, Springer Biographies,
https://doi.org/10.1007/978-3-030-95033-0_22

L Lift of aircraft
L' Lateral moment about the aircraft's center line
q_∞ Dynamic pressure
ρ Air density
V_∞ Free stream velocity relative to aircraft

1   Introduction

In 1889 Otto Lilienthal published his book *Birdflight as the Basis of Aviation* [1] describing his experiments proving the concept of cambered wings and providing the best available instructions of that time on how to design a winged aircraft. Starting in 1891, Otto Lilienthal made the first, documented, successful flights on a heavier-than-air man-carrying aircraft under the control of a human being. In 1893, he patented the world's first production aircraft, the *Normalsegelapparat*, a monoplane glider of which he sold at least nine machines to various places in Europe and America. He performed more than 2000 successful flights leading to his aerodynamic data and flight reports being circulated around the world. Samuel Langley as well as Octave Chanute and many others corresponded with Lilienthal. Chanute, author of the other most influential book of that time [2] and later a friend of the Wright brothers, followed Lilienthal's approach to carefully perform flight tests and with that led several younger men to successful flight performances in 1896 and beyond. Wilbur Wright wrote in his article [3] about Otto Lilienthal: *"… he was without question the greatest of the precursors, and the world owes to him a great debt."* According to the patent description. the monoplane glider falls under the category covered by the FAR Part 103, which is described in more detail in the *Weight-shift Control Aircraft Flying Handbook* (FAA-H-8083-5) of the *Federal Aviation Administration* [4]. Weight-shift control aircraft are still being flown, safely and successfully, by many thousands of pilots around the world. However, after Otto Lilienthal's fatal accident in 1896 very little has been known about later attempts to fly this aircraft.

2 Glider Dimensions During Wind Tunnel and Flight Tests

Wind tunnel tests gave new insights into the performance, trim state, flight stability, and the influence of the permeability of the fabric, which has been woven on an original loom using a formula that was developed based on a careful analysis of fabric taken from an original glider wing. Two parameters that depend on the weight, size, and fitness of the test pilot are the wing area and the pilot's position in the glider. Their selection was motivated by the analysis of the wind tunnel data presented in the following. However, in order to focus on the findings gained during the flight tests, only the two diagrams, which helped determining the best possible glider dimensions, will be discussed here (Table 1).

Based on the commonly used classification of Lilienthal's gliders [5], the *Normalsegelapparat* (*Normal Soaring Apparatus*) has a wingspan of up to 7 m and a maximum chord of 2.5 m. According to Lilienthal's descriptions, larger dimensions lead to aircraft that are difficult to control in wind gusts. The wind tunnel model used in the 2017 full-scale wind tunnel tests at the largest European wind tunnel, the DNW-LLF, had a span width of 6.7 m and a surface area of 13.2 m^2. Its main frame location differed slightly from the patent drawing [6] by some centimeters, which has been found to be of great importance to the pilot's required posture for stable flight. A free stream velocity of 11.5 m/s and an aircraft geometric angle of attack of $\alpha = 5.82°$ ($\alpha = 0°$ corresponds to a vertical main frame orientation) leads to sufficient lift to lift the glider (25 kg) plus a person of Otto Lilienthal's weight ($\triangleq$ 80 kg). (See Fig. 1).

It was discovered during the force and moment measurements that the glider built for the wind tunnel experiments was difficult to trim. It tended to be tail heavy, even for a small negative angle η of the horizontal tail plane. Additional tethered flights in the wind tunnel confirmed these findings. During the wind tunnel run, the test pilot could not continuously hold his body in the most frontal position because of the approximately one-minute time required to accelerate the wind tunnel flow. In contrast to the glider's orientation going downhill, in the wind tunnel, the glider at the free

Table 1 Comparison of Lilienthal with test pilot

	Otto Lilienthal	Test pilot
Overall weight	80 kg	90 kg
Body height	180 m	1.92 m

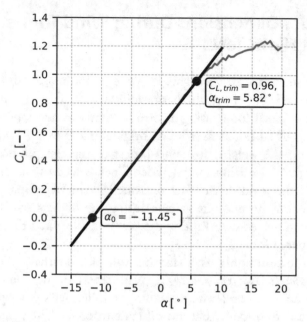

Fig. 1 Lift coefficient versus geometric angle of attack measured using the wind tunnel glider at 11.5 m/s. (▶ https://doi.org/10.1007/000-6z1)

stream velocity is inclined upwards by approximately 15°, which made the pilot's most forward leaning posture even more arduous. The tethered flight in the DNW-LLF confirmed that an additional 8.2 kg trim weight at the front of the cockpit was required to fly the wind tunnel glider version stably in a horizontal setting. This nose down force was applied through a ballast cord at the cockpit front pulling downwards (see Fig. 2).

The larger weight and body size of the test pilot led to the following dimensions of the outdoor glider, which later was used for flight testing: the wing span was chosen to be 7 m and the maximum chord length 2.5 m, resulting in a wing surface area of 15.6 m². The fact that the main frame location with respect to the wing area center differs on the wind tunnel model and the patent drawings (see Fig. 3), and that the wind tunnel model required a ballast force to reach steady horizontal tethered flight, led to a main frame position of the outdoor glider that was 134 mm further ahead of that of the wind tunnel glider. The main frame position limits the pilot's front position and therefore influences trim state and static margin and/or the posture required to move the center of gravity in front of the aerodynamic center, which is necessary to obtain a stable and trimmed flight. It is well known that weight-shift control aircraft tend to handle best with small static margins for obvious reasons. However, small static margins do also imply that just

Fig. 2 Steady tethered flight with an externally applied nose down moment correction, corresponding to an 8.2 kg trim ballast at the aircraft's front.
(▶ https://doi.org/10.1007/000-6yx)

a few centimeters displacement of the relative center of gravity changes the handling qualities of the aircraft from 'easy to control' to 'impossible to fly'.

Wing surface geometries and sizes at identical center positions of the three aircraft versions are depicted in Fig. 3. Instead of the aerodynamic center, the wing's area center has been chosen as the reference point for the three glider versions, since the aerodynamic center was measured only for the wind tunnel version of glider. However, it is assumed that wing surface center and aerodynamic center position are displaced at very similar quantities for all three glider versions because of their geometric similarity.

Following the wind tunnel experiments, measurements of the longitudinal center of gravity location of the wind tunnel glider were performed for each of the pilot's various postures. The evaluation of these data, together with the force and moment data, showed the same deficit of the static margin that was present on the wind tunnel glider. The necessity of the trim ballast can be explained by the fact that the main frame location, and therefore the pilot's center of gravity, differed between the wind tunnel model and the patent drawings depicted in Fig. 3 (top). A detailed analysis of a vertical projection of the glider wings and the location of the glider main frame showed a distance between the wing surface area center and the main frame at the patented glider of 452 mm, compared to only 348 mm for the wind tunnel model (see Table 2). A correction to the pilot position within the glider was applied to the measured data and the results showed that the patented glider could be flown stably at moderate negative tail plane angles without ballast given a pilot posture as depicted in Fig. 4c (trimmed).

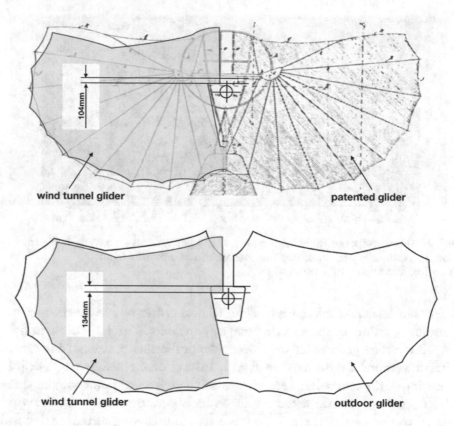

Fig. 3 Comparison of wing surface geometry and size at identical main frame positions of three aircraft

Table 2 Comparison of the three glider versions

	Wind tunnel glider	Outdoor glider	Patented glider
Overall weight	25 kg	32 kg	20 kg
Relative pilot position	348 mm	452 mm	482 mm
Wing area	13.24 m^2	15.60 m^2	14.13 m^2
Static margin	< 0	57.7 mm	56.4 mm
Stability (no ballast)	✗	✓	✓

3   Familiarization of the Test Pilot—Tethered Flights and Winch Supported Flights

The first thing the test pilot had to learn when attempting to fly the monoplane glider was longitudinal and lateral control by weight shifting.

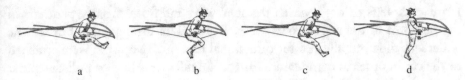

Fig. 4 Pilot's postures during start **a**, flight (**b** climb and **c** trimmed posture) and landing **d**. (▸ https://doi.org/10.1007/000-6yy)

Directional control is only performed indirectly by lateral control. Otto Lilienthal gave the following advice for longitudinal control [2]: "*The first rule is to keep your legs well extended toward the front, and in landing to throw the upper part of the body backward, so that the front edge raises itself and thus checks the motion, as may be seen whenever a crow alights.*" This leads to the most forward center of gravity position during takeoff and the most backward position during landing. They are described by the sketches of postures shown in Fig. 4, which were drawn according to the figures given in the patent drawing by Otto Lilienthal.

Lilienthal advises to always use the following procedure to start [2]: "*The starting and landing must be done exactly against the wind. The fixed vertical rudder will keep the apparatus exactly in the wind when in a state of rest. The horizontal rudder keeps the apparatus from tipping over forward, a thing that arched surfaces are inclined to do.*" In our case, it was decided to start with tethered flights on a $5 \times 5 \text{ m}^2$ platform towed by a passenger car. This allowed for frequent training at reduced risk and costs. The glider as well as the pilot was attached to the platform corners with safety bonds (pilot) and wires (glider) as depicted in Fig. 5. The pilot was wearing a harness connecting him to four safety bonds. The glider had eyebolts anchored to the lower ends of the main

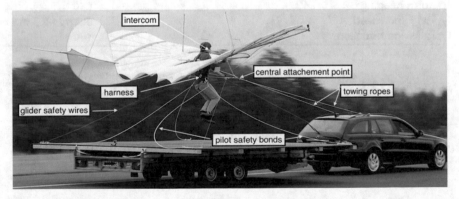

Fig. 5 Practicing at "zero ground speed," limited height and limited control authority—platform tests. (▸ https://doi.org/10.1007/000-6yz)

frame cross bars through which the steel wires ran from the front corners of the platform to the rear corners on each side. This arrangement limited the glider's altitude but allowed for directional motion. Two ropes were attached to the car in order to compensate for the aircraft's drag and to pull the glider forward. Given a glider altitude less than half a meter above the platform and a lateral position in the center of the platform, the ropes attached to the platform corners were loose. The pilot could therefore test the influence of weight shifting in longitudinal direction, leading to the postures depicted in Fig. 4.

The two pull ropes were attached to the ends of the glider main frame cross bar by another four ropes, which intersected at a central attachment point 0.7 m in front of the main frame. Therefore, the longitudinal stability was improved and the pilot postures had a reduced influence on the pitching motion.

Heading and wind directions did not match during the tethered flight tests because they were conducted on an airfield with just one runway. Therefore, the pilot had to compensate for a cross-wind component of up to 5 m/s (Table 3 and Fig. 5).

The experiences gathered during the tethered flights on the platform were required for the next phase of winch supported tests. The winch tests had to be conducted without the safety wires, which during the platform tests not only limited the flight altitude, but also the roll motion of the glider. The

Table 3 Platform flights

June 2017	
Distance in air	230 m–400 m
Platform speed	9.3 m/s–9.8 m/s
Platform heading	060
Typical wind speed	3.9 m/s–5.3 m/s
Typical wind direction	140

Fig. 6 Practicing lateral control with limited longitudinal control authority—winch tests. (► https://doi.org/10.1007/000-6z0)

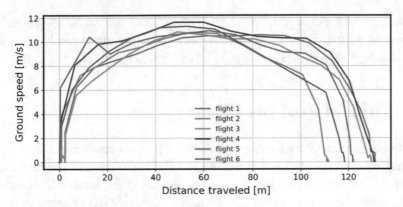

Fig. 7 Winch speed versus distance traveled

Table 4 Winch flights

October 2017	
Distance in air	98 m–220 m
Altitude while airborne	−0.1 m–4.3 m
Speed while airborne	8.7 m/s–11.8 m/s
Rope force	390 N–1140 N
Glider heading (block position)	240
Average wind speed	0.7 m/s–1.3 m/s
Average wind direction	240

pilot's lateral control skills became the key issue in cases of moderate gusts on the one hand, and also in cases of stall[1] that can lead to asymmetric lift on the other hand. Lilienthal described the procedure of lateral control and related difficulties as follows [2]: "*The following mistake is to be particularly avoided. The experimenter is soaring in the air and feels himself suddenly raised by the wind, but unequally, as is usually the case—for instance, the left wing more than the right. The inclined position forces him toward the right. The beginner involuntarily stretches his legs to the right, because he foresees that he will strike the earth on the right hand. The result is that the right wing, which is already lower, is loaded still more, and flight tends more and more downward and to the right, until the tips of the right wing strike the earth and are broken. For life and limb there is less danger, as the apparatus forms an efficient guard in every direction, which checks the force of the blow. The correct thing to do is always to extend one's legs toward the wing that is rising, and thus to press it down again.*

---

[1] Asymmetric lift during stall occurs especially at low pitch rates Q or initial high attack angles. The influence of the leading edges vortices inherent to dynamic stall at flare landings will be described later.

In the beginning this requires some force of will, but this useful movement soon becomes an unconscious one, after we see how surely the wings can be guided this way and be protected from damage." (Table 4 and Figs. 6 and 7).

4   Trimming of the Glider

The winch tests in Germany were performed with an attachment point using a carabiner clasp again placed approximately 0.7 m in front of the main frame center. This additional restoring moment increased the longitudinal stability considerably. Thus, the longitudinal control authority was reduced as long as the pulling rope was under tension. The fact that the outdoor glider was still a bit tail heavy had been overlooked, as the pitch-up tendency became relevant only at low rope tensions when the glider was in descent and the pitch-up was intended and supported by the pilot's landing posture (Fig. 4d). During the free flight in California though, this needed to be corrected by either a ballast weight or a reduction of wing area ahead of the pilot. The first measure chosen to increase the static margin for a given stabilizer angle and center of gravity was the reduction of wing area in front of the glider by a v-shaped enlargement of the cockpit 'window', the gap in the fabric in front of the pilot. Additionally, after some remaining problems with longitudinal control and stability, the wing area at the glider's front was further reduced. The front ribs of the wings were moved a few centimeters backwards and the fabric along the leading edge of the wing was cut and re-attached accordingly.[2] The resulting final shape of the glider is depicted in Fig. 3.

Figure 8 depicts $C_{M,cg}$ versus a_0 for three different pilot postures for the final version of the outdoor glider. The data have been derived from the wind tunnel force and moment data measured in 2017 making the following assumptions: 1. The lift increases proportionally with the wing area (15.6/13.24); 2. The weight of the pilot (90 kg) is located further forward than during the wind tunnel tests, because of the altered main frame position of the outdoor glider used for flight tests (134 mm). The dotted and dashed line in Fig. 8 corresponds to the climb position (Fig. 4b), the dashed line to the trimmed position (Fig. 4c), and the solid line to the start position (Fig. 4a). It should be noted that the dashed line intersects with the horizontal axis for $C_{M,cg} = 0$ at $a_0 = 17.25°$. This defines the trim angle of attack and the horizontal tail plane angle and the pilot posture during trimmed, stable, and controllable downhill flights at 11.5 m/s when the systems weight, drag,

---

[2] In one case Lilienthal attached some fabric at the rear part of the glider (Kleiner Doppeldecker).

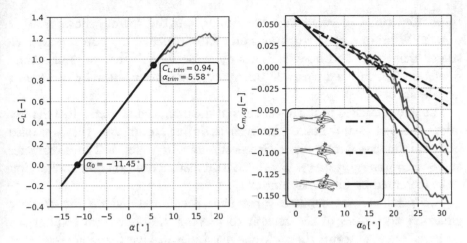

Fig. 8 Lift coefficients versus geometric angle of attack derived for the outdoor glider at 11.5 m/s (left). Moment coefficient versus absolute angle of attack derived for the outdoor glider (right). (▶ https://doi.org/10.1007/000-6yw)

and lift forces are in equilibrium. The three necessary criteria for longitudinal balance and static stability, namely positive moment coefficient at $\alpha_0 = 0$, negative moment curve slope, and trim angle of attack within the flight range, are fulfilled. The measured data correspond very well to the conditions found during the downhill flights, after the start, but before the landing.

After this modification, the glider could easily be controlled in the longitudinal and lateral directions as well as in free downhill flight, as in cases of flights that were initially accelerated featuring a bungee cord. The latter method was used when wind and slope of the terrain were too weak for pure foot launching. The bungee cord was a well-suited means to increase the available practicing time and made it possible to get airborne at Dockweiler Beach in the vicinity of Los Angeles and at Tres Pinos in the vicinity of Hollister (CA). Due to the fact that this bungee cord was attached to one central hook directly on the main frame, the trim, stability, and control authority was the same with and without the bungee cord. All final flights were performed freely without any ropes attached at Marina Beach north of Monterey (CA).

The outdoor glider used in the end for the free flights differed in some other minor aspects from the gliders used by Otto Lilienthal and others at his time.

The outdoor glider has been coated with white wood glue. This glue behaves physically just like other types of collodion that was used to coat Lilienthal gliders. After it is applied to human skin, for example, it acts like a sort of second skin in both cases. The monoplane glider was folded many times, sometimes in very rough conditions, but the coating shows no sign of wear. Wind tunnel investigations on the influence of the porosity demonstrated that forces and flow fields are the same for different ways of sealing the fabric. So, this is a chemical difference, but aerodynamically of little or no relevance.

The skids attached to the lower ends of the main frame were—like the test pilot's kneepads—only in use after mishaps during landing. So, they were also of little influence during flight. Stainless steel wire ropes of 2 mm diameter and oval compression sleeves have been used instead of steel wires and tension locks (patented by Lilienthal himself).

Two parallel short bamboo sticks have been mounted on the outdoor glider on both sides of the cockpit to prevent the pilot from leaning too far back. They substitute the backrests that Lilienthal used, and are—like the helmet—a further concession to todays safety considerations. Luckily, neither the helmet nor the bamboo sticks have ever been needed.

The arm rests of our outdoor glider were made more primitively. The pilot could lean on them like Lilienthal on his, but he couldn't lift the glider in the back in order to get the minimum incidence angle for the start at low-wind conditions. During the winch flights and the final downhill flights this was not necessary because the head wind lifted the tail immediately after the first few meters. At low wind conditions, the pilot used shoulder straps from time to time to lift the tail before start and thereby facilitate the acceleration.

A structural difference between our outdoor glider and the version which was described by Lilienthal in his patents and publications is described in the following. Willow was used only for parts that had to be bent significantly, like the cockpit frame and the vertical and horizontal stabilizer. However, pine wood sticks, which had been bent into shape while being soaked in water for three days, were used instead of willow for the struts of the wings. This increased the weight to 32 kg and might have shifted the center of gravity of the glider a bit towards the rear. Still, it needed to be done due to the test pilot's weight being approximately 15% greater than Lilienthal's and the required lift forces are approximately another 15% higher when flying horizontally (tethered and at the winch rope) than during downhill flights. As for Lilienthal's gliders, bamboo was used only when parts of a certain length had to be stiff and straight, like the tail.

5   Acquirement of Longitudinal Control Skills—Free Flights

When practicing on a training hill at Tes Pinos in the vicinity of Hollister (CA), the wind tended to be gusty. During long periods over the year, the winds blow from north-west along the valley, but on the test day they were dominated by a strong upper western wind leading to flow separation at the western California coast ranges, which created a few occasional rapid changes in local wind strength and direction. A sudden gust in the early flight phase after takeoff created an upward acceleration, which made keeping the pilot's posture for trimmed condition impossible. The altitude variation could later be reconstructed by correlation-based image processing from the videos (see Fig. 9). A 10th-order polynomial was used to fit the altitude data estimated for each 0.1 s and to derive the rate of climb (or descent respectively) and the load factor for a time interval of 3.5 s (Fig. 10).

The additional inertial forces made it impossible to keep the legs directed straight forward. The glider was lifted approximately 4 m within one second and the glider's incidence angle moved from approximately 15° downwards to 50° nose up. As a consequence, the pilot's weight moved backwards within the cockpit. Lilienthal routinely jumped from a roof of more than 4 m high on routine basis and managed acrobatically to pitch the nose down and avoid roll, even in those vulnerable stalled conditions without lowering his legs. Lilienthal experimented with an actively steered tail plane, flaps and wing

Fig. 9 Stall and post-stall behavior. (▶ https://doi.org/10.1007/000-6z2)

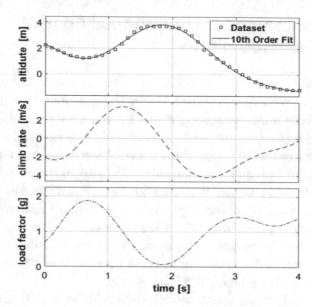

Fig. 10 Altitude, rate of climb/descent, and load factor during the stall event depicted in Fig. 9

Fig. 11 Preparation, start, and initial flight phase with lateral and longitudinal control. (► https://doi.org/10.1007/000-6z3)

warping, but intentionally decided to stay with the weight shifting control as this allowed him to operate the glider reliably during thousands of flights [7]. The many flights that he started by jumping off the building roof at the *Fliegestation* near Steglitz (a suburb of Berlin, Germany) required the start of the glider reliably at zero ground speed. It is hard to imagine that this can be done better using active control means, and thus weight shift control is still the method of choice for most unpowered foot launched aircraft. However, foot launching gliders in gusty wind conditions requires a great deal of practice. As a consequence of the pilot's inability to counteract stall, in Tres Pinos the glider stopped and fell approximately four meters like a parachute: the tail plane flipped upwards, while no strong nose down moment occurred, and the wings stayed nearly level. This safety feature of the glider was described in Lilienthal's US-Patent [6] with the words: "... *On the latter is pivoted the tail q in such a manner that it can freely turn upward, but finds downward a point of support on the fixed rudder r. This mode of attaching the tail has the advantage that the tail will have no carrying action when the machine is employed like an ordinary parachute, thereby preventing from turning downward.*" It should be noted that the tail couldn't transfer an upward force, because it is rigidly coupled with the upper ropes when the horizontal stabilizer acts downwards, but flips up when a force acts on it upwards. The upwards flexibility prevents the tail from breaking when it touches the ground during flare landings. It is well understood that these observations cannot substitute an analysis of dynamic stability of the aircraft. However, they indicate that extreme situations can exceed the pilot's capacity to maintain his posture required for balanced, controlled flight. Due to the pivoted tail plane, stall seems to be less problematic when a pilot behaves cautiously and does not counteract stall in a later stage. Lilienthal's annual flight reports not only informed others about successful flights, but also about occasional failures. One such a flight accident described in detail was a rapid dive that occurred after a small change of wing curvature. This can be interpreted in a way that these modifications had an effect on the static stability of the glider. Additionally, it seems likely from the authors' experience that the required pilot posture to flare, as depicted in Fig. 4d, will potentially exceed the pilot's strength in such a situation. The dimensions of the glider used for the free flights, corrected geometric imperfections of the wind tunnel model and increased the static margin to an amount that matched the static margin of the patented glider. The glider has been safely flown after the described incident, but strong wind gusts were more consequently avoided.

The final pitch trim was obtained by a moderate negative angle of attack of the horizontal tail plane. As a result, the glider reacted nicely and sensitively to the pilot's input and could easily be directed against the wind. During several starts on the sand dune, shear winds required counteracting a descending left wing counterintuitively by shifting the pilot's weight below the right wing. Here, the training during the winch-supported flights paid off, and when the wind lowered one of the wings the pilot shifted his legs to the other side in an instinctive manner (see Fig. 11). Basically, the control of the roll angle has to be performed like the one of a modern hang glider, but the legs need to travel greater distance side-wards in order to create a similar reaction of the glider, due to the smaller weight that is shifted, when repositioning just the legs. In addition to the fact that a modern glider is laterally controlled by the relocation of the pilot's whole weight, they also amplify the roll moment L' by shifting the keel and therefore warp the wings accordingly [4]. It has to be noted that turns cannot safely be executed while flying at low altitude in the vicinity of a hill. Therefore, the question of whether steep turns can safely be performed remains unanswered by our tests. It is known from Lilienthal's flight reports that he—who flew much higher—made turns, but (still) tried to avoid them since a safe landing has to be conducted against the wind.

Lilienthal reported that landing requires a similar counter-intuitive move as is the case with turning the glider. He reported that the pilot has to move his legs backwards to pitch up and decelerate, even if his instinct advises him to have his feet in the front when approaching the ground at higher speeds. However, this depends on the trim of the glider and in the case of our test flights with the monoplane glider, it was just enough to lean backwards with the upper body and therefore move the weight of the whole body to the rear (see Fig. 12). The only problems that occurred while coordinating the landings at the beginning were the same problems beginners face during their first hang glider lessons. When initiating the landing too early and too slowly, stall

Fig. 12 Straight, controlled down-hill flight and landing. (▶ https://doi.org/10.1007/000-6z4)

occurs in a way that the flow on the wings separates slowly but massively. As massively separated flow is never steady nor two-dimensional, one wing starts sinking earlier than the other and generates more drag at the same time. This leads to an unintended turn at the end of the flight with both, Lilienthal's gliders as well as with any modern weight-shift control aircraft. The trick to landing the Lilienthal glider well is doing this maneuver a little later and at higher pitching rates Q, so that the stall occurs dynamically. The dynamic stall vortices along the leading edges of the wings will then force the flow into a two-dimensional condition [8], while creating a short lift overshoot and an additional pitch-up moment [9]. After these lessons were learned, the glider could be flown reliably and steadily for up to 70+ meters and the coordination of gentle straight flare landings were performed routinely.

6   Conclusion and Outlook

Otto Lilienthal's patented monoplane glider (*Normalsegelapparat*) was flown reliably and steadily downhill over distances up to 72 m. The static margin of the glider as laid out in the patent drawings of Lilienthal can safely be flown by a person with a weight of not more than 80 kg, who is capable of holding his legs straight forward, even under the influence of gravitational and inertial forces during maneuvers. Strong wind gusts have to be avoided. The dihedral of the wings and the static margin of the glider, which is dominated by the main frame location with respect to the aerodynamic center, can be increased to help inexperienced pilots to encounter unintended pitch and roll. Seven-meter wing spans are no problem for tall pilots. Sealing the wings, for example with a balanced wood glue water solution, is advised. Pine wood can be used for the wing ribs instead of willow in order to facilitate the manufacturing, but leads to a heavier aircraft. In addition to that, pine wood is less flexible and does not allow for an optimization of the three-dimensional shape by simply adjusting the wires below the wings. A comparison of Lilienthal's *Normalsegelapparat* from 1894 and its replica from 2018 (Fig. 12) suggests that Lilienthal's glider had less dihedral, a stronger wing curvature in the center, and a larger wing span with respect to the body size of the pilot (see Fig. 13).

Preliminary flight tests with the *Large Doppeldecker* (see Fig. 14), the bi-plane, which was flown by Lilienthal successfully from 1895 to 1896, showed that the stability of the bi-plane is higher, but that flight speed and agility are larger in the monoplane. Further tests of the aerodynamic performance and the mechanical strength of the bi-plane are ongoing.

Fig. 13 Top: Otto Lilienthal and Paul Beylich—*Fliegeberg* near Berlin [10]. Below: Markus Raffel and Andrew Beem—Sand City near Monterey. (▶ https://doi.org/10.1007/000-6z5)

Fig. 14 Preliminary flight tests with Lilienthal's large bi-plane (*Großer Doppeldecker*)

Acknowledgements Markus Krebs' help during the glider manufacturing and the experiments is greatly appreciated. He and Felix Wienke took part in the experiments in Germany. Their skills and dedication were the key to success of the whole project. The staff of the *Otto-Lilienthal-Museum* in Anklam (Germany) built the bi-plane and gave advice during the manufacturing of the monoplane, which has been made of authentic materials after Otto Lilienthal's patent drawing. JT Heineck supported Markus Raffel's flight tests not only by making his truck available for three months, but as a real friend in many other ways, too. Andrew Beem (*Windsports*, LA) is likely the best American hang glider flight instructor. He gave extremely valuable input and support during the experiments in California. Gerhard Fahnenbruck's and Arno Trümper's initiation of the project is greatly appreciated. Many thanks to Simine Short, biographer of Octave Chanute, and the *Otto-Lilienthal-Museum*. The team of DNW's *Large Low Speed Facility* (LLF) helped passionate and competent during all phases of the wind tunnel tests. The project was strongly supported by many flight enthusiasts of the *German Aerospace Center* DLR.

Appendix 4: On the Flying Qualities of Otto Lilienthal's Large Biplane

Markus Raffel, Felix Wienke, and Andreas Dillmann

Nomenclature

A_f	Flight altitude ($A_f = 0 \triangleq$ to straight legs; feet on ground)
α	Aircraft's angle of attack ($\alpha = 0$ zero lift)
C_D	Drag coefficient of aircraft
C_L	Lift coefficient of aircraft
C_M	Longitudinal moment coefficient about aerodynamic center
C_P	Pressure coefficient
c	Chord length
E	Glide ratio C_L/C_D
η	Geometric tail plane angle of attack
Q	Pitch rate
q_∞	Dynamic pressure
ρ	Air density
Δs	Flight speed
U_∞	Free stream velocity
v_f	Flight speed.

Supplementary Information The online version contains supplementary material available at (https://doi.org/10.1007/978-3-030-95033-0_23). The videos can be accessed individually by clicking the DOI link in the accompanying figure caption or by scanning this link with the SN More Media App.

M. Raffel and B. Lukasch, *The Flying Man*, Springer Biographies,
https://doi.org/10.1007/978-3-030-95033-0_23

1   Introduction

More than 125 years ago the aviation pioneer Otto Lilienthal was the first person to invent, build, and publicly fly several aircraft. He received a US patent for his monoplane glider [1] in 1895. Several copies of this *Normal Soaring Apparatus* were sold to customers in America and Europe. In the same year, he developed his designs further into two different biplane aircrafts, of which the *Large Biplane (Großer Doppeldecker)* showed the most promise. Lilienthal's idea behind the transition from his monoplane design to the biplane depicted in Fig. 1 was to increase the wing surface without enlarging the wingspan, as this would have made controlling the aircraft in roll more difficult. Countless flights with both biplanes have been photographically documented, making them the first successful, man-carrying biplanes in history. Lilienthal's flight demonstrations and his theory of cambered wings, developed and published in his book [2], contributed to the epochal shift in the rapid development of aeronautics. Culick [3] also notes that he was the first aeronautical engineer to combine the accepted concepts of equilibrium and stability with his ideas of control in order to maintain equilibrium in the face of disturbances. Amongst other experts and flight enthusiasts, the

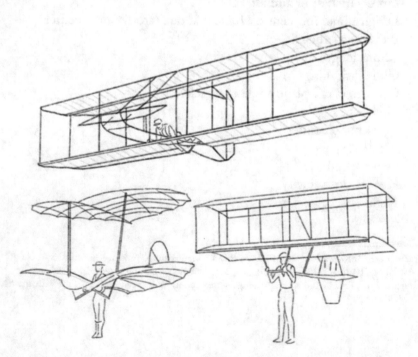

Fig. 1 1902 Wright glider (top) and its precursors: 1895 Lilienthal's *Large Biplane* (left) and 1896/97 Chanute-Herring glider (right)

American railroad engineer Octave Chanute corresponded with Lilienthal. According to Crouch [4], Chanute served as the focal point of the international community of aviation pioneers at the time by corresponding with leaders of the field such as Lilienthal and Langley. He supported the cause of aviation by spreading news, holding lectures and establishing a baseline of shared knowledge through publications such as his classic book on flying machines [5]. In the period between 1894 and 1904 Chanute decided to begin his own experiments. By following Lilienthal's approach of carefully performing increasingly advanced flight tests, he guided several young men, amongst them Augustus Herring, towards successful flight performances. He introduced bridge building techniques to the truss structures of his bi- and multiplanes to improve their structural integrity as shown in Fig. 1. In his experiments in 1896, this biplane flew as stable as Lilienthal's biplane [6] with a larger wing span and improved structural rigidity. Chanute focused on maintaining equilibrium in flight by incorporating automatic stability in his designs, which led him to make first steps towards active controls.

Finally, the Wright brothers combined the existing body of experience and knowledge with their own innovations in the field of active pilot controls and aerodynamics in their extensive glider tests between 1900 and 1902. Applying Lilienthal's step-by-step approach, they were able to achieve the first powered flights in late 1903. Actually, acquiring pilot skills, before attempting powered flight, made the "airmen" so much more successful than the preceding attempts of the "chauffeurs". Culick [7] states that most of the Wrights' predecessors focused on intrinsically stable aircraft and did not progress far enough to be concerned with the problem of maneuverability. According to Perkins [8], the Wrights believed from the beginning that powerful controls were mandatory and would allow the pilot to maintain the necessary equilibrium. Their breakthrough became possible, because their designs exhibited reduced stability complemented by reasonably effective pilot controls around all three spatial axes.

The present work is intended to give insights into the aerodynamic and handling properties of Lilienthal's large biplane. The goal is to further the understanding of Lilienthal's achievements as one of the greatest of the precursors, as Jakab [9] calls him. It is a continuation of the authors' investigation into its monoplane predecessor presented in Refs. [10] and [11].

The *AIAA 1903 Wright Flyer Project* [12] pursued similar goals of constructing and testing a full-scale model as well as performing manned flights with a minimally modified replica of the historic Wright flyer. Further research such as virtual reality simulations based on the test data followed [13]. Another investigation into the Wrights' 1901 and 1902 gliders was published by Kochersberger et al. [14, 15], who evaluated the aerodynamic performance from full-scale wind tunnel tests. They were able to derive simulation models, which were used in preparation of manned flights in a replica of the 1902 glider. Lawrence and Padfield performed further investigations into the handling qualities of the Wrights' 1902 glider [16] and powered 1905 flyer 3 [17]. They derived simulation models from their reduced scale wind tunnel data to evaluate their flight dynamics behavior and maneuverability. A thorough discussion of the handling qualities was based on piloted simulation trials. They conclude that the flight control system was the Wright brothers' most important innovation in their early development, followed by continuous improvements towards the powered 1905 flyer.

2   Glider Reconstruction and Dimensions

Earlier DLR wind tunnel tests demonstrated the influence of the permeability of the fabric, which has been woven on an original loom using a formula that was developed based on a careful analysis of a fabric sample taken from an original glider wing [10]. The lower wing of the *Large Biplane* is the exact same size as the one used in Lilienthal's patented monoplane glider. Unlike the lower wing, the upper one is not foldable but is divided in the middle. Besides increasing lift, the upper wing also changes the flight mechanical properties in comparison to the monoplane, which will be discussed below. The original *Large Biplane* glider did not survive. However, there are several preserved specimens of the original monoplane, on which the biplane was based. The authentic replica of the *Large Biplane* used for the tests described in this paper was built by the *Otto-Lilienthal-Museum* in Anklam (Germany). It is reported that Lilienthal modified his gliders to a certain extent during his experiments [18]. In conjunction with the wood and fabric construction reinforced by steel wires, it can be assumed that the glider geometries were subject to various alterations. The geometry of the replica (main dimensions given in Table 1) is based on surviving drawings by Lilienthal [1], as well as on drawings by Nitsch [18]. Circular arc airfoils were manufactured with a

Table 1 Glider dimensions

Length	5.25 m (17 ft 3 in)
Wing span	6.60 m (21 ft 8 in)
Wing area	24 m^2 (259 sq ft)
Empty weight	33.5 kg (74 lb)
Mean aerodynamic chord length	2.03 m (6 ft 8 in)

camber to chord ratio of 1/20, which is in line with the camber ratios documented by Lilienthal [7, p. 271]. Special attention was paid to the tension of the steel wires connecting the willow longerons of the wings to the mainframe as depicted in Fig. 2. Their lengths greatly influence the overall trim of the glider. During all balance measurements and later flight tests, the wing fabric was sealed with a coating of diluted wood glue. This treatment resulted in a flexible coating, which was easier to apply than the collodion coating originally used by Lilienthal. Using glue instead of collodion for coating has no aerodynamic consequences because the remaining permeability to air of both treatments is negligible.

There are some minor differences between the replica and Lilienthal's *Large Biplane*, which are intended to reduce the pilot's risk during the flight tests. To prevent the mainframe from digging into the ground during landing mishaps, skids were fitted to the ends of the mainframe. Their aerodynamic effect is negligible due to their small size. The original solid wires, bracing the wing longerons against the mainframe, were replaced by stainless steel cables. Their tension has been permanently fixed using compression sleeves instead of adjusting it using the custom tension locks, which Lilienthal patented for his gliders. The backrests were replaced by two parallel bamboo rods along the sides of the cockpit for safety reasons. They provide the pilot with a larger range of motion, while preventing him from leaning too far backwards, allowing him to place some of his weight on them. The tube-shaped forearm supports of the original were reduced to sturdier, flat arm pads, located on the lower mainframe struts. The pilot used these as arm rests during flight, but he did not have any leverage to lift the rear of the glider during takeoff. During the final downhill flights, the head wind was always strong enough

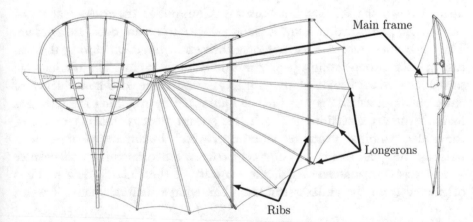

Fig. 2 Basic structure of the biplane's mainframe and lower wing (The ribs consisted of rails defining the profile of the wing.)

to lift the whole glider by itself after a short takeoff run of only a few meters. When low wind conditions persisted during the towed flights, the arm pads were supplemented with shoulder straps that bore most of the glider's weight, making the takeoff run less strenuous. The strongly curved parts of the glider structure, such as the cockpit frame, stabilizers, and longerons, were manufactured from willow, just as in the original. However, the three longerons closest to the leading edge were replaced by pine wood poles, which were presoaked and bent into shape. The improved structural stability at the cost of an increased glider weight of 33.5 kg was necessary since one of the test pilots was about 15% heavier than Lilienthal. In addition, tethered and towed horizontal flights added another 15% to the required lift forces when compared to free downhill flights at an incline.

3   Balance Measurements

The full-scale balance tests of the *Large Biplane* were conducted with the help of the test vehicle of the *German Hang Gliding Association* (DHV) in Fürstenfeldbruck (Germany) as depicted in Fig. 3. A three-component balance was mounted on a sting at the upper end of a supporting tower structure on top of the vehicle and clamped to the glider's mainframe.

Data was recorded at free stream velocities of $20 \, \text{km/h} \leq U_\infty \leq 45 \, \text{km/h}$ at angles of attack between $-17° \leq \alpha \leq 45°$. The influence of the pilot's drag was estimated by a drag penalty based on earlier tests, assuming the steady flight body posture shown in Fig. 4b. The elevator incidence angle was set to the middle position of $\eta_2 = -22.5°$ of the three calibrated angles depicted in Fig. 4e. All data was recorded during measurement runs up and down the 2.7 km long runway. Continuous traverses in angle of attack were performed at a low angular velocity to obtain quasi-steady data. The test vehicle records free stream velocity and direction during the test runs to take current atmospheric conditions into account. The results were sorted into discrete angle of attack intervals and averaged. The influence of atmospheric disturbances was further minimized by repeating the test runs in both runway directions for each free stream velocity. During the tests, the glider exhibited structural vibrations, which manifested themselves as noise in the quasi-steady measurement curves. Since the sting and balance system was comparatively rigid, the majority of these elastic deformations originated from the glider's structure. The wing structure, made of wood

Fig. 3 Balance force and moment measurements with the Lilienthal glider mounted on the supporting structure of the test vehicle of the *German Hang Gliding Association* (DHV)

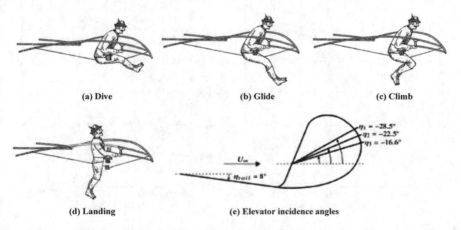

(a) Dive   (b) Glide   (c) Climb

(d) Landing   (e) Elevator incidence angles

Fig. 4 Pilot postures and elevator incidence angles

and steel bracing wires, proved itself to be a well-designed truss structure, exhibiting small and continuous deformations in the form of wing bending. The tailplane structure was connected to the mainframe by a single bamboo rod of approximately 30 mm in diameter and braced with cotton chords against the main structure. This resulted in vertical up- and down bending and longitudinal torsion of the whole tailplane structure during testing. The tailplane vibrations were strongest at high angles of attack, when they were triggered by turbulence from the stalled wing. Changes in the elevator incidence angle due to deformation of the tail structure under the aerodynamic load have not been measured. The digital filter described by Savitzky and Golay [20] was applied to the averaged forces and moments to reduce the small-scale variations introduced by noise and vibration.

The measured lift coefficients over the angle of attack are shown in Fig. 5 at several free stream velocities. Up to an angle of attack of $\alpha = 12°$, all lift curves exhibit a linear interval with differences in the order of $\Delta C_L = 0.12$ between the highest and the lowest free stream velocity. With increased velocity, the lift tends to increase at a given angle of attack. The stall behavior is benign, with a gradual departure from the lift curve and low gradients near the maximum lift coefficient between $1.10 \leq C_L \leq 1.25$ at an angle of attack of approximately $\alpha \approx 20°$. The drag polar shown in Fig. 5 has a classical parabolic shape in the linear lift interval below $C_L = 1.1$. The minimum drag of $0.123 \leq C_{D\,min} \leq 0.155$ occurs near $C_L \approx 0.25$. An increase in free stream velocity results in a reduction of the minimum drag.

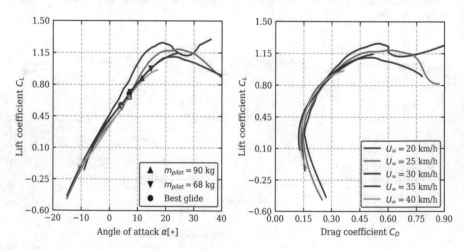

Fig. 5 Lift coefficients plotted against angle of attack (left) and drag coefficient (right) at different free stream velocities

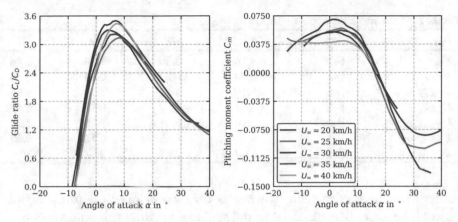

Fig. 6 Glide ratio (left) and pitching moment coefficients (right) at different free stream velocities

The pitching moments around the glider's mainframe center are shown on the right-hand side of Fig. 6. They exhibit three distinct intervals. The first interval up to an angle of attack of $\alpha = 10°$ coincides with the linear lift interval. All free stream velocities exhibit positive, pitch-up pitching moments around $C_M = 0.055$, with a tendency towards smaller values at higher velocity. The change in pitching moment in the linear interval is too small to confirm a linear dependency on the angle of attack due to the measurement uncertainty. Between $10° \leq \alpha \leq 25°$, the glider transitions into full stall with a linear decrease in pitching moment and little variation between the free stream velocities. Above $\alpha = 25°$, the glider is fully stalled and the pitching moment tapers off. Lower free stream velocities result in more negative, pitch-down moments. The glide ratio $E = C_L/C_D$ is shown on the left side of Fig. 6. The glider achieves a maximum glide ratio of $E = 3.5$ at a free stream velocity of $U_\infty = 35$ km/h. The lowest glide ratio of $E = 3.1$ was measured at a free stream velocity of $U_\infty = 25$ km/h.

Based on the weights of the two test pilots and the glider, trim conditions are calculated for the five measured free stream velocities. They are listed in Table 2, along with the conditions for best glide and maximum lift, and included as markers in Fig. 5 for free stream velocities $U_\infty \geq 30$ km/h. The required free stream velocity for flight at $C_{L,\max}$ with a given pilot weight determines the minimum takeoff velocity $U_{\infty,\min}$. The lighter pilot (68 kg) is able to take off at velocities above 28 km/h, while the heavier pilot (90 kg) requires at least 31 km/h. The best glide ratio is achieved at lift coefficients between $0.57 \leq C_L \leq 0.7$ at velocities above 30 km/h. The lighter pilot is able to fly the glider in its best glide state at 35 km/h, which is only 25%

Table 2 Trim and best glide conditions

U_∞	[km/h]	25	30	35	40
E_{opt}	[–]	3.14	3.30	3.50	3.44
$C_{L,opt}$	[–]	0.74	0.57	0.70	0.66
$C_{L,max}$	[–]	1.19	1.11	1.14	0.96
$C_{L,trim,90kg}$	[–]	–	–	0.87	0.67
$U_{\infty,min,90kg}$	[km/h]	30.0	31.1	30.6	33.4
$C_{L,trim,68kg}$	[–]	–	0.98	0.72	0.55
$U_{\infty,min,68kg}$	[km/h]	27.2	28.2	27.7	30.3

faster than the takeoff velocity, whereas the heavier pilot has to fly slightly faster. The glider operates close to the stall region with both pilot weights.

4   Acquiring Lateral Control Skills—Winch Flights

Initial experiments quickly showed that the lateral control skills of the pilot are the deciding factor for keeping prolonged flights safely level because of external disturbances such as gusts. Asymmetric lift occurred especially at low pitch rates or initial high angles of attack as depicted in Fig. 7. The influence of the leading edges' vortices inherent to dynamic stall at flare landings will be described later. Lilienthal himself describes the pilot input, which counteracts roll disturbances, as a lateral shift towards the rising wing in [21]. The measured data of the winch supported flights is summarized in Table 3.

Table 3 Winch flights—August 2018

Distance flown Δs	119–380 m
Altitude while airborne A_f	−0.1–3.8 m
Speed while airborne v_f	7.5–8.9 m/s
Average wind speed u_∞	0.5–1.9 m/s

Fig. 7 Practicing lateral control with limited longitudinal control authority due the attachment of the towing rope in front (a) and counteracting unintended roll during winch tests (b) (▶ https://doi.org/10.1007/000-6z8)

5   Acquiring Longitudinal Control Skills—Free Flights

After practicing lateral control near Moringen (Germany), Marina Beach (CA), and on the dunes near Kitty Hawk (NC), the *Large Biplane* replica has now been flown by three pilots: Markus Raffel (DLR), Andrew Beem (*Windsports*), and Billy Vaughn (*Kitty Hawk* Kites). Foot launching gliders in gusty wind conditions requires some practice. However, by acquiring those skills, it was eventually possible to fly the glider safely. Strong wind gusts were consequently avoided because it was assumed that they can lead to a stall, which can exceed the pilot's capability to maintain the posture required for a balanced, controlled flight.

The angle of incidence of the horizontal tail plane was adjusted along the three angles shown in Fig. 4e to achieve longitudinal trim for a given pilot mass. A steeper angle of $\eta_2 = -22.5°$ was suitable for the heaviest pilot (M.

Raffel) at 90 kg, while the lightest pilot (Andrew Beem) at 68 kg achieved the best flights at a shallower angle of $\eta_3 = -16.6°$. The glider responded promptly and predictably to the pilot's input in these configurations. As a result, the pilots were able to easily direct the glider against the wind. This proved valuable during takeoffs from a sand dune shown, as shown in Fig. 8a. If one wing descends (here for example the left wing), the intuitive reaction of an untrained pilot is to also shift the legs to the left in order to land safely on his feet. However, since the torso is fixed in position with respect to the Lilienthal glider, this motion moves the center of gravity to the left, which amplifies the leftwards roll angle. This can result in a flip of the aircraft,

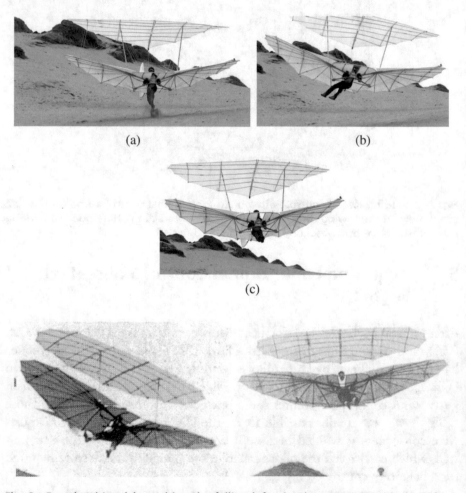

(a)   (b)

(c)

Fig. 8 Foot lauching (a), catching the falling left wing by counterintuitively shifting the legs toward the right wing (b), and establishing straight flight (c) (top: A. Beem, below: O. Lilienthal) (▶ https://doi.org/10.1007/000-6z7)

potentially causing a dangerous crash with the arms stuck in the framework of the glider. The correct, but counterintuitive, response is to shift the legs towards the rising wing. This is visible in the takeoff from the sand dune (Fig. 8a), where shear winds initially pushed the left wing down. After the training in Germany and California the pilot instinctively shifted his legs towards the rising wing, leveling the glider (Fig. 8b).

The lateral control of the roll angle is similar to a modern hang glider, with two important differences. With a modern glider, the entire pilot mass is shifted relative to the wing. In addition, the roll moment is amplified by warping the wing through a shift of the wing keel [22]. With the historic glider, only the legs can be laterally repositioned. As a consequence of the smaller shifted mass, the legs need to be moved further to the side to achieve a sufficient response by the glider. Due to the comparatively low control effectiveness in roll, it was deemed unsafe to perform turns at low altitude and in close proximity to a hillside. Therefore, the feasibility of steep turns could not be investigated during the free flight experiments, although Lilienthal demonstrated 180° turns at higher altitudes. However, even Lilienthal tried to avoid such maneuvers because he asserted that safe landings could only be conducted against the wind.

After steady downhill flight depicted in Fig. 9, the pilot initiated the landing phase by shifting his weight rearwards to pitch the glider up and slow it down when it comes close to the ground. The biplane glider exhibited a similar sensitivity to longitudinal weight shifts as the monoplane so that the pilot only had to lean his torso backwards, as shown in Fig. 10. Premature and tentative weight shift results in a slow but pronounced stall of the flow over the wings. Since massively separated flow is always unsteady and three-dimensional in nature, the glider rolls towards the wing, which stalls first and creates less lift and more drag.

To achieve a good landing of Lilienthal's biplane, the pitch rotation needed to be delayed and the pitch rate increased to have the wings stall dynamically. The result was a short lift overshoot coupled with an additional pitch-up moment [23], which was caused by the two-dimensional flow over the wings due to dynamic stall vortices along their leading edges [24, 25]. With these lessons learned, Otto Lilienthal's 125-year-old feat of landing the glider on one foot was reproduced by Andrew Beem, as depicted in Fig. 10b. Different

Fig. 9 Lift-off (**a**), acceleration (**b**) and initial flight phase with lateral and longitudinal control (**c**) (top: A. Beem, below: O. Lilienthal)

pilots were able to perform smooth and safe flights followed by gentle landings in the replica, thereby achieving flight durations of up to 15 s over distances given in Table 4.

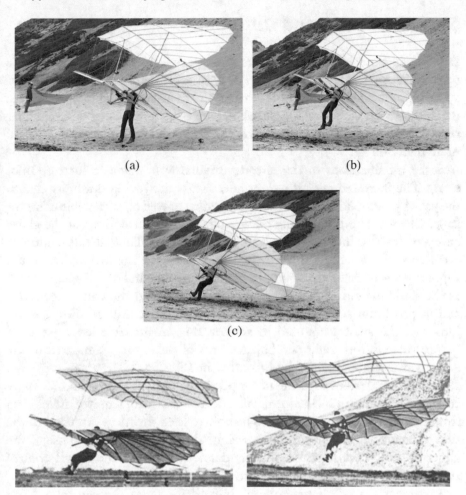

Fig. 10 Initial weight shift (**a**), pitch up and deceleration (**b**), and controlled one foot landing (**c**) (top: A. Beem, below: O. Lilienthal) (▶ https://doi.org/10.1007/000-6z6)

Table 4 Free flights—July 2019

Distance flown Δs	20–104 m
Altitude while airborne A_f	0.1–5.5 m
Speed while airborne v_f	4.5–7.6 m/s
Average wind speed u_∞	1.8–6.9 m/s

6   Conclusion and Outlook

Free flights from hillsides as well as winch tows close to the ground were performed using an authentic replica of Otto Lilienthal's *Large Biplane* glider, achieving free flight distances of up to 104 m. The structural integrity of the design has been verified through balance measurements using a hang glider test vehicle, which demonstrated that the glider can lift and sustain loads of more than 240 kg.

Some modifications to the historic original were made to increase pilot safety. The increased takeoff weight and the reduced wing flexibility due to the use of pine wood for some of the longerons proved an acceptable compromise. Changing and improving the various adjustment options of the glider gave new insights into the likely learning process of Lilienthal and reinforced our respect for his ingenuity. For example, changing the inclination of the support posts of the upper wing can increase the dihedral of the wings as well as the static margin of the glider [26]. This improved the control behavior and helped less experienced pilots to cope with pitch and roll disturbances. However, the alignments used to achieve this, might have led to reduced performance during flight and balance tests of the replica, compared to the flight performance obtained by Lilienthal in 1895/96.

Longitudinal trim was achieved for pilot weights between 68 and 90 kg by adjusting the elevator incidence angle. The lateral control required athleticism from the pilots because they had to be able to keep their legs extended to the side to have control input through weight shift. This limited the use of the glider to wind conditions of only moderate gusts since the lateral control authority was deemed too low for strong wind gusts.

A comparison of photographs of Lilienthal's *Large Biplane* from 1895 (Fig. 11) to its replica from 2018 (Fig. 12) suggests that Lilienthal's original glider had a little less dihedral, a stronger wing curvature in the center of the lower wing, and less wing washout on the upper wing. The aerodynamic center of Lilienthal's *Large Biplane* is higher than that of his patented monoplane, increasing the flight stability while reducing the lateral control authority. Therefore, the biplane's stability is greater, but flight speed and agility are higher in the monoplane.

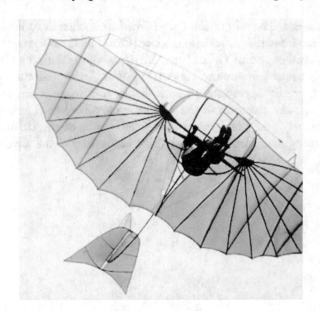

Fig. 11 1896 Otto Lilienthal—*Fliegeberg* near Berlin [19]

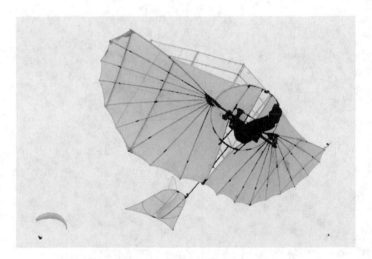

Fig. 12 2019 Andrew Beem—Marina Beach near Monterey
(▶ https://doi.org/10.1007/000-6z9)

Acknowledgements The staff of the *Otto-Lilienthal-Museum* in Anklam (Germany) built the biplane from authentic materials after Otto Lilienthal's patent drawing and photographs. Andrew Beem (*Windsports*, LA) flew the biplane in California and gave extremely valuable input and support. Markus Krebs' help during the winch experiments is greatly appreciated. Many thanks to Simine Short, biographer of Octave Chanute, and the team of the *Otto-Lilienthal-Museum*. The team of DHV helped passionately and competently during all phases of the balance tests. The project was strongly supported by many flight enthusiasts of the *German Aerospace Center* DLR.

Appendix 5: Wind Tunnel and Flight Testing of Otto Lilienthal's Experimental Monoplane from 1895

Markus Raffel[iD]**, Pascal Weinhold, Felix Wienke, Clemens Schwarz, and Andreas Dillmann**

Nomenclature

A_f Flight altitude ($A_f = 0 \triangleq$ to straight legs; feet on ground)
α Aircraft angle of attack ($\alpha = 0$ zero lift)
c_D Aircraft drag coefficient
c_L Aircraft lift coefficient
c_l Aircraft rolling moment coefficient
c_m Aircraft pitching moment coefficient
c_n Aircraft yawing moment coefficient
E Glide ratio C_L/C_D
η Geometric tail plane angle of incidence
Θ Rudder deflection angle
κ Opening angle of the leading edge flaps
U_∞ Free stream velocity.

Supplementary Information The online version contains supplementary material available at (https://doi.org/10.1007/978-3-030-95033-0_24). The videos can be accessed individually by clicking the DOI link in the accompanying figure caption or by scanning this link with the SN More Media App.

M. Raffel and B. Lukasch, *The Flying Man*, Springer Biographies, https://doi.org/10.1007/978-3-030-95033-0_24

1   Introduction

In 1889 Otto Lilienthal published his book *Birdflight as the Basis of Aviation* containing the first lift-versus-drag data of cambered wings and other important information required for human flight [1]. In 1895 he build and flew an experimental monoplane, which featured already various control surfaces that became so important for later pioneers, but not much is known about his efforts to publish and share his findings in public. At that time, many of his more general flight experiences were regularly detailed in articles that were published for example in Germany, France, Russia and the United States [2, 3]. Many people from around the world came to visit him, including Russian Nikolai Zhukovsky, Englishman Percy Pilcher and Austrian Wilhelm Kress. Zhukovsky wrote that Lilienthal's flying machine was the most important invention in the aviation field. Lilienthal corresponded with many members of the *Boston Aeronautical Society*—of which he was an honorary member—among them Octave Chanute, author of *Progress in Flying Machines* [4], James Means, who invited him to come for flight demonstrations, Samuel Pierpont Langley who visited Lilienthal in Berlin, and Greely S. Curtis, who even gained first-hand experience in gliding on a visit with Lilienthal beginning in September 1895.

On August 5, 1895, Otto Lilienthal wrote to Octave Chanute in Chicago that he had been able to significantly improve his flying machine: the new mechanical system made the device so stable in flight that anyone could easily learn its use. The German *Imperial Patent Office* in Berlin issued patent no. 84417 for this innovation as a supplement to Lilienthal's earlier aircraft patent from September 1893. The patent applied to a "*Form of the flying machine protected by patent no. 77916 in which the front part of the wing surface can be rotated downwards around the leading edge and pressed downwards by elastic elements in such a way that it rotates downwards when the air pressure ceases to act from below, thereby creating a moment that lifts the apparatus.*" The drawing attached to the patent is schematic, illustrating the patent claim without giving many details. Since the patent was issued as a supplement, the elevator (horizontal stabilizer) is depicted in front of the fin (vertical stabilizer) as in 1893, even though Lilienthal had positioned both stabilizers in a crosswise configuration at the end of the tail for quite some time at that point [5] (Figs. 1 and 2).

Figure 3 depicts a very large monoplane, with the leading edge formed by a flap. This leading edge flap gave the apparatus the name *Vorflügelapparat (Front Wing Apparatus)* in some later German literature. Lilienthal, however, called it the *Experimental Apparatus* [6]. In order to follow Lilienthal's naming

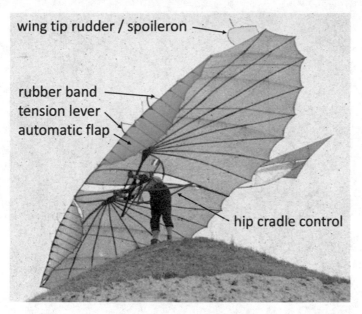

wing tip rudder / spoileron

rubber band
tension lever
automatic flap

hip cradle control

Fig. 1   Otto Lilienthal in his *Experimental Monoplane* on top of his training hill *Fliegeberg*, 1895. Photo: P. W. Preobrazhensky, 1895

and to distinguish the apparatus from others, we will call it the *Experimental Monoplane* in the remainder of this article.

The leading edge flap however, should not be confused with a "slat" i.e., a slat wing in today's sense. The purpose of this leading edge flap, in contrast to a modern slat which has its trailing edge above the leading edge of the main wing, is not to prevent or reduce stall at high angles of attack. On the contrary, the effect of Lilienthal's flap was a positive pitching moment to be achieved at low or even negative angles of attack. Otto Lilienthal called flow acting on the leading edge of the wings from above the *Oberwind*, *(Upper Wind* or *Top Wind)*. This describes an effect that primarily, but not only, occurs during launching, when the pilot has to pick up speed at approximately zero angle of attack and thus the lowest possible drag. Time and again, vexatious falls occurred during launching especially the monoplanes, both to Lilienthal and during the flight tests with his replicas described earlier [7]. If the leading edge of a glider is held slightly too low, the wind suddenly presses it down, the pilot can no longer hold the apparatus with its front pointing upwards and falls. Lilienthal wanted to counter this danger of pitching down at low angles of attack during launching and flight with the flexible flap. One of his drawings suggests that the flap was originally intended to be operated manually, however the device was then pretensioned by rubber bands to be

OTTO LILIENTHAL in BERLIN.

Flugapparat.

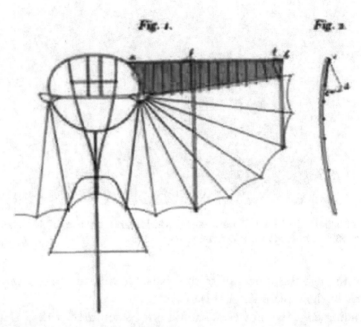

Fig. 2 Illustration from Lilienthal's second flying machine patent no. 84417, pending dated May 29, 1895

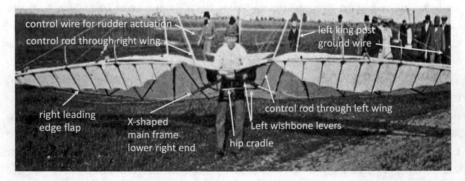

Fig. 3 Paul Beylich in the *Experimental Monoplane* "with automatic balancing". Hip cradle and the control levers and rods transferring the control input to the strings on the Wings can clearly be seen. *Photo* R. Neuhauss

Fig. 4 Otto Lilienthal in his *Experimental Monoplane* flying at his training hill *Fliegeberg*, 1895. *Photo* R. Neuhauss

open at rest so that the flap would close automatically when sufficient airflow from below the flap was present.

The *Experimental Monoplane* (see Figs. 4 and 5) was designed for control experiments from the very beginning. The actuating system can already be seen in the photo (Fig. 1) of May 29, 1895, taken by the Russian photographer P. W. Preobrazhensky [8]. It consisted essentially of the hip cradle formed by two willow withies bent upward toward the rear and joined at the end near the vertical stabilizer. They were connected by an additional bar in front of the pilot's body. With wishbone-like levers connecting the hip cradle and control rods on each side, a displacement of the hip fork will cause deflection of the control rods, which pass through the wing surface near the rear cockpit ring. In the photos, these oblique control rods look like two additional king posts.[1] At first, they were only used to deflect the vertical stabilizer laterally as a rudder. The movement of the control rods was transferred to the trailing edge of the tail via cords [5]. These cords ran from the king posts over the control rods to the lower part of the tailplane. It is not clear from the photos whether the flexibility of the tail boom or an articulated

---

[1] In contrast to modern hang gliders, which have one king post and ground wires to prevent the wind bending downwards while being on the ground, Lilienthal's monoplanes usually had two vertical posts for the same purpose.

Fig. 5 Paul Beilich (shouldering most of the load) and Otto Lilienthal (helping at the leading-edge) bringing the *Experimental Monoplane* with hip cradle, spoilerons and actuated rudder back to the starting position on top of Lilienthal's training hill (*Fliegeberg*) close to Berlin. *Photo* P. W. Preobrazhensky, 1895

pivot provided the lateral deflection of the tail unit's trailing edge. However, an articulated arrangement with only moderate rudder deflection long lever arms to the pivot point of the tail boom is more likely in the author's view (Figs. 6, 7 and 8).

Figure 4 shows the lateral deflection of the rudder while Otto Lilienthal pushes the hip cradle to the left. This also results in a center of gravity shift which causes a roll moment, whereas the resulting rudder deflection in the same direction produces a yaw moment and thus allowed for coordinated turns. In order to increase the effect of the rudder, Lilienthal had

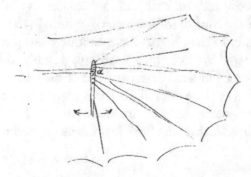

Fig. 6 Illustration of Lilienthal's wing warping mechanism as sketched and described in a letter to Wolfmüller

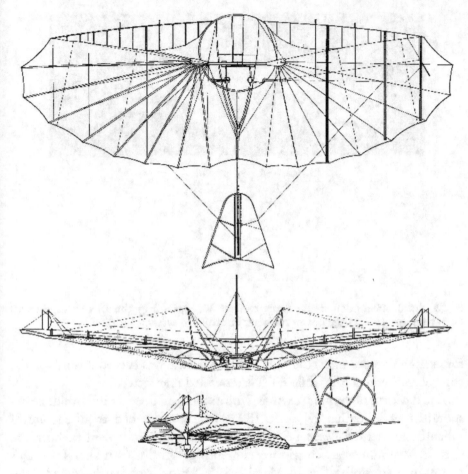

Fig. 7 Reconstruction drawing of the *Experimental Monoplane*

enlarged the vertical stabilizer by adding ribs and fabric to the top. However, the size of the elevator stayed the same as for his *Normal Soaring Apparatus* resulting in reduced longitudinal stability of the aircraft with closed leading edge flaps, which made the newly invented automatic pitch control even more important.

Another interesting control experiment is documented by the photos of the Russian photographer P. W. Preobrazhensky in the summer of 1895. Small vertical control surfaces were attached to the outer ends of the wings, which could rotate around a short post like small sails. A string led from the front edge of such a wing tip rudder or roll spoiler to the control rod, which had been shortened and moved back for this purpose. In normal flight, these control surfaces align themselves in the wind. If the pilot moves his body,

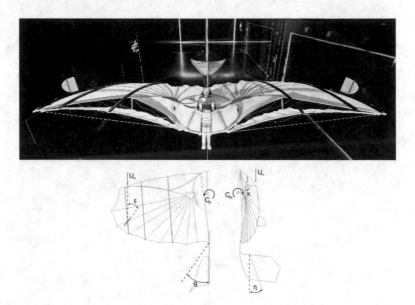

Fig. 8 (top) 1:5 model of the *Experimental Monoplane* in the DLR-SWG. (below) Deflection angles of the control surfaces; coordinate system

for example, to the right, the right control surface was turned inward via the control rods and cords, while the left remained unaffected.

The Bavarian flight pioneer Alois Wolfmüller had been an important correspondent of Lilienthal [9] since 1893. In 1894, he had acquired one of Lilienthal's patented *Normal Soaring Apparatus* and made flight experiments with it. Wolfmüller also began experimenting with his own aircraft designs to improve controllability. In March 1895, Otto Lilienthal informed Alois Wolfmüller of his large *Experimental Monoplane*: *"I am currently building a larger glider of about 20 m² wing area, which can only be used at calm winds."* He wrote about the wing tip rudders: *"Furthermore, I have attached a surface to each wingtip, which I can straighten up by pulling the string in order to bring back the leading wingtip."* Since he wrote of *"straighten up the surface"*, it is possible that, in addition to the devices visible in the photo, there were experiments with simpler spoilerons. This control device is interesting in that it has the potential to avoid the problem of adverse yaw. If such a control surface is deflected to one side by moving the hip cradle, the drag is simultaneously increased on that wing and the lift is decreased. As an actuated spoiler increases the drag, the yaw is in the same direction as the roll [10]. This asymmetric actuation of aircraft spoilers is still used by airliner pilots today, allowing the aircraft designers to install smaller ailerons. This technique is used, especially during descending flight, when the drag increase is welcome to reduce altitude.

It is also noteworthy that both Wolfmüller and Lilienthal worked with warping of the wings. Wolfmüller wanted to use this control method manually, but Lilienthal most likely connected it with the hip cradle as did the Wright brothers in their 1902 glider and the 1903 flyer. In August 1895, Lilienthal wrote to Wolfmüller: "*You are entirely correct. The shift in center of gravity must be greater than a person can accomplish, when gliding in the wind with large wings. As the simplest method of balancing the lifting capacity of the two wings, I recommend rotating the wings around the longitudinal axis. I have found this to be the safest method compared to any others. It is also the method that is used by birds.*" After exchanging assurances of "*mutual agreement to the protection of legitimate interests*", a kind of non-disclosure agreement, Wolfmüller presented his thoughts and the results of his experiments on controlling flying machines built using Lilienthal's design in a long letter dated September 28, 1895. This prompted Lilienthal to write more freely about his own attempts. Wolfmüller had designed a wing warping device, as well as an installation that would allow pilots to sit within the flying apparatus. He argued that a sitting position would be advantageous, freeing up the pilot's hands to operate mechanical control systems such as two levers for twisting the wings in his own design. Other control elements could be operated using a strap around the upper body.

The same year in October, Lilienthal replied: "*I tested an arrangement similar to yours for moving and rotating the wings with outer tensioning wires running to different points of a lever mounted at the lower base point that can be pulled to give the wing profile the desired rotation. I also made it so that the tail could rotate to the right or left, making it easier to land. Furthermore, I have attached a surface to each wingtip, which I can straighten up by pulling the string in order to bring back the leading wingtip. These elements were operated by the hips, which press against a hip cradle, when the body is shifted sideways to shift the center of gravity.*" Lilienthal concluded by admitting that he had not yet achieved a decisive breakthrough in controllability: "*These experiments, which I spent the entire summer investigating, have prompted me to make significant changes that I have not yet fully clarified and for which I regrettably have little time at the moment.*"

The full-scale replica as well as the 1:5 model built for this investigation had a complete set of controls: rubber-band activated leading edge flaps for automatic pitch control, and spoilerons, wing warping and rudder for yaw and roll control, which were actuated by a hip cradle individually or combined. All structural materials relevant to the flying qualities, were selected with great care in order to match the characteristics of the original.

2   Wind Tunnel Tests

A range of parameter sets were investigated with the 1:5 model of Otto Lilienthal's *Experimental Monoplane* in two different wind tunnels. The main focus of the investigation was on the aerodynamic effects of the various control elements. Isolated deflections of elevator, rudder, wing warping, spoilerons and leading edge flaps were compared to a reference configuration with undeflected control elements. The main part of the aerodynamic investigations was carried out in the DLR-SWG, which is a closed-loop, low-speed wind tunnel. A closed test section with a length of 9 m, a width of 2.4 m and a height of 1.6 m was used. At the maximum power of $P = 0.5$ MW a maximum flow velocity of $U_\infty = 65$ m/s can be achieved in the empty test section. The lack of cooling requires an active flow velocity control system, which reduces the variations of the Reynolds number resulting from temperature changes. Each configuration was examined for up to 15 different angles of attack and at three mean flow velocities of $U_\infty = 5$ m/s, 7 m/s, and 8.5 m/s. It was not possible to investigate higher velocities in a safe manner due to insufficient structural stability of the model. Wind tunnel effects on angle of attack, as well as lift and drag coefficients and the pitching moments were corrected using classical linear methods. The measurement system consisted of a 6-component RUAG 796-6C strain gauge balance, a Prandtl tube, a Hottinger and Baldwin MGC_{plus} measurement amplifier system, and a computer network.

For the investigation of the wing warping, spoileron effects and hinge moments at the leading edge flaps, an additional experiment in DLR's 1-m low-speed wind tunnel (1MG) was conducted. The right half of the existing 1:5-model was mounted directly on a piezoelectric force- and moment balance and exposed to the flow through the wall of the wind tunnel. Yaw and roll moments, as well as hinge moments of the leading edge flaps were recorded for a single flow velocity of $U_\infty = 8.5$ m/s (Table 1).

2.1   Performance

The lift over drag polar of the glider is depicted in Fig. 9. The approximately quadratic shape with an offset towards positive lift coefficients is characteristic of a cambered wing. The glider enters the stalled flow regime for lift

Table 1 Test matrix

Flow velocity in $\frac{m}{s}$	5	7	8.5
Rudder deflection Θ_0 in °	0	0	0
Rudder deflection Θ_1 in °	3.7	2.2	1.8
Rudder deflection Θ_2 in °	7.3	5.9	5.4
Rudder deflection Θ_3 in °	11.4	10.0	9.1
Elevator inclination η_0 in °	−11	−10.5	−9.5
Elevator inclination η_1 in °	−2.5	−2.0	−1.8
Elevator inclination η_2 in °	−22.8	−21.3	−19.6
Wing warping ww_0	±0	±0	±0
Wing warping ww_1	Neg	Neg	Neg
Spoileron deflection ϵ_0 in °	0	0	0
Spoileron deflection ϵ_1 in °	90	90	90
Leading edge flap deflection $\varkappa_0$ in °	0	0	0
Leading edge flap deflection $\varkappa_1$ in °	30	30	30

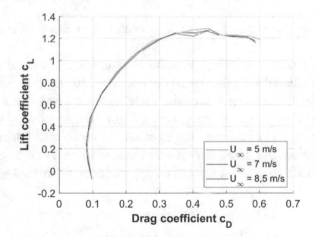

Fig. 9 Lift versus drag of the *Experimental Monoplane* with closed leading edge flaps

coefficients above $c_L = 1.1$ and achieves a maximum lift coefficient of $c_L = 1.25$ at an angle of attack of $\alpha = 22.3°$. A minimum drag coefficient of $c_D = 0.078$ was recorded. The influence of the free stream velocity is negligible, which indicates a minor Reynolds-dependency of the results. It also suggests that the structural deformations are relatively small, since the shape of the wing does not change with the increasing dynamic pressure. Figure 10 shows the glide ratio E between lift and drag as a function of the angle of attack. The glide ratio forms a distinct maximum between $6.8° \leq \alpha \leq 9.2°$. Due to the limited number of measured angles of attack, the maximum glide ratio and the angle of attack at best glide can only be determined approximately.

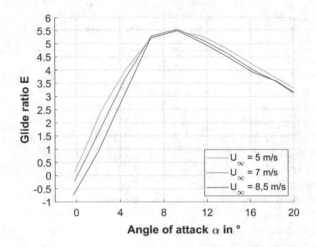

Fig. 10 Glide ratio of the *Experimental Monoplane* with closed leading edge flaps

The maximum glide ratio $E_{\max} = 5.55$ occurs at an angle of attack of $\alpha = 9.3°$.

In order to assess the flying qualities in manned flight conditions, trims points are calculated for pilot masses of $m_{\text{pilot}} = 70$, 80 and 90 kg using a lift curve, which is averaged over the three measured free stream velocities. The resulting trim points for an assumed flight velocity of $U_{\infty} = 11.5$ m/s are shown in Fig. 11. The trim angles of attack are located between $9° \leq \alpha \leq 12°$ well below the onset of stall and close to the maximum glide conditions. Lilienthal's reported weight of 80 kg results in a trimmed glide ratio of

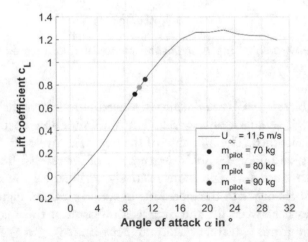

Fig. 11 Lift coefficient versus aerodynamic angle of attack for three different trim points

$E_{Trim} = 5.3$ at an angle of attack of $\alpha_{Trim} = 10.25°$, which is only about 5% below the best glide value. A previous investigation by Wienke et al. [11] on Lilienthal's first patented production aircraft, the *Normal Soaring Apparatus*, arrived at a trimmed angle of attack of $\alpha = 16°$ at a significantly lower glide ratio below 4 for the same pilot mass and flight velocity. In comparison, the best glide ratio of the *Experimental Monoplane* is about 40% higher, which is due to the considerably larger wingspan at identical pilot drag and horizontal stabilizer dimensions. It is questionable whether some imperfections like the pilot dummy, which was slightly too small and without clothes, led to a bias towards higher values of the glide ratio. It is understood that the full-scale tests performed with Lilienthal replicas earlier delivered a higher accuracy. However, the main advantage of the *Experimental Monoplane* is its increased wing surface, which allows the 80 kg pilot to fly very close to the best glide ratio.

2.2   Stability

Since not all early experimental aircraft designs were stable with respect to their flight mechanics, the static longitudinal stability characteristics are now discussed based on the measured pitching moment curves. Several conditions have to be met in order to achieve steady, trimmed and statically stable flight. The total mass of glider and pilot along with the flight velocity result in a trim angle of attack on the lift curve, which has to fall in the range of attached flow below maximum lift. At this trim angle of attack, the location of the combined center of gravity and the elevator incidence angle have to be chosen such that the pitching moment around the combined center of gravity becomes zero. This flight condition is statically stable, when the slope of the pitching moment curve around the combined center is negative as it crosses the $c_m = 0$-abscissa from positive to negative pitching moments (Fig. 12).

The present glider is controlled through pilot weight shifts and changes of the elevator incidence angle. Figure 13 shows pitching moment curves around the main frame center location for the three investigated elevator incidence angles. An increasingly negative elevator incidence angle η shifts the pitching moment curve to larger values. As a result, the elevator could be used as a trim device before takeoff.

The pitching moment results are now compared to data previously published by Wienke [12] for Lilienthal's preceding aircraft, the patented *Normal Soaring Apparatus*, in order to give them a better context. Figure 12 shows the pitching moment curves of the patented glider around four different trimmed centers of gravity at a flight velocity of $U_\infty = 11.5$ m/s.

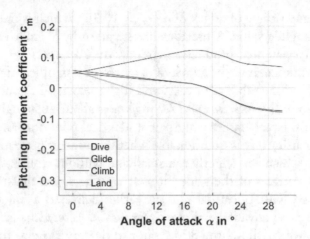

Fig. 12 Pitching moment coefficients around trimmed centers of gravity of the *Normal Soaring Apparatus* for four different pilot postures [12]

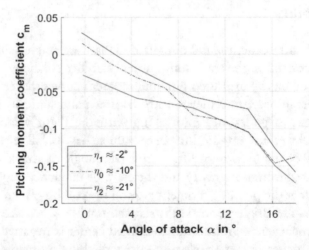

Fig. 13 Pitching moment coefficient around the main frame center of the *Experimental Monoplane* for three different elevator inclinations

The effect of the weight shift on the linear pitching moment curves below α < 16°, illustrate the effect of a center of gravity shift. As the center of gravity shifts backward from dive to glide position, the slope of the pitching moment curve increases. Figure 14 compares the pitching moments around the glide center of gravity for the *Normal Soaring Apparatus (Normalsegelapparat)* with the results of the *Experimental Monoplane* at its most negative elevator incidence angle for both open and closed leading edge flaps. The results of the present study are shown as linear approximations, which were derived from

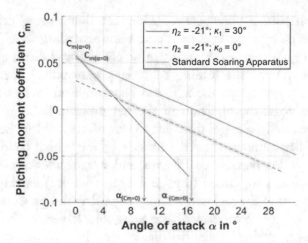

Fig. 14 Linearized pitching moment coefficient around the center of gravity (pilot posture glide) of *Normal Soaring Apparatus* and *Experimental Monoplane* for two different settings of the leading edge flap (at maximum elevator inclination)

measured data of the attached flow region below $\alpha < 20°$, in order to suppress measurement noise that occurred when large portions of the flow separated to a larger extent.

The pitching moment curves of the two configurations are similar, but exhibit different slopes and intersections with ordinate and abscissa. First, the zero-lift pitching moment coefficient c_{m0} at the zero-lift angle of attack of $\alpha = 0°$ will be discussed. The zero-lift pitching moment $c_{m0} \approx 0.05$ of the *Normal Soaring Apparatus* is significantly higher than the $c_{m0} \approx 0.03$ of the *Experimental Monoplane* at closed leading edge flaps $\kappa_0 = 0$. It follows from this lower zero-lift pitching moment that the pitching moment coefficients of the *Experimental Monoplane* are below those of the *Normal Soaring Apparatus* for the whole angle of attack range up to the trim angle of attack α_{Trim}. It can also be deduced that the slope of the pitching moment coefficient at the trim angle of attack is lower, which indicates less static stability. The zero-lift pitching moment $c_{m0} \approx 0.05$ of the *Experimental Monoplane* with open leading edge flaps $\kappa_1 = 30°$ is significantly higher and coincides with that of the *Normal Soaring Apparatus*. This indicates similar statically stable flight characteristics for both aircraft in this configuration. Because of the more negative pitching moment slope of the fully open leading edge flaps, the trim angle of attack for a given center of gravity is also smaller in the open configuration. For the main frame position considered for the data analysis shown in Fig. 14, the trim angle of attack reduces from about $\alpha_{Trim} = 17°$ down to $\alpha_{Trim} = 7°$. In order to fly the glider with open leading edge flaps, the pilot would have to shift the center of gravity too far to the front for a sustainable pilot posture. However, flying the glider with open flaps is not necessary because of their automatic actuation, which is explained next. Figure 15 illustrates the working principle of Otto Lilienthal's automatic pitch control system. The leading edge flaps were pulled open by rubber bands, whose tension could be adjusted before takeoff. When the angle of attack was reduced, the direction of the net pressure force on the leading edge flaps eventually changes from lift to downforce, which then opens the leading edge flaps supported by the tension of the rubber bands. Figure 16 depicts color coded pressure coefficients and stream lines derived from 2d-CFD computations for closed (top) and open (bottom) leading edge flaps. It can be seen, that for a closed leading edge flap the pressure is continuously higher on the lower side of the wing and lower on the upper side. This indicates a relatively equal lift distribution in cord-wise direction. For an open leading edge flap, the flow field shows a strongly increased lifting pressure difference on the leading edge flap and a reduced lift force on the

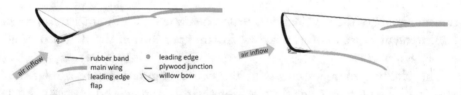

Fig. 15 Closed (left) and open (right) leading edge flap positions depending of the incoming flow's angle of attack

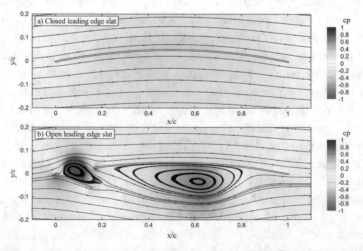

Fig. 16 Color coded pressure coefficients and stream lines derived from 2d-CFD computations for closed (top) and open (bottom) leading edge flaps

main wing due to the reduced pressure values on the main wing's lower side. The pitching moment is therefore considerably higher for the open leading edge flap, at the price of a reduced overall lift and an increased drag.

The described consequences of the different flow fields can also be found in the measured wind tunnel data and can be seen in the linear trend of the pitching moment curve as depicted in Fig. 14. This likely extends below the zero-lift angle of attack $\alpha = 0°$ so that an open leading edge flap shows to produce a higher, nose-up pitching moment than the closed, baseline configuration. As a result, the opening of the leading edge adds a returning moment towards positive angles of attack. The beauty of Otto Lilienthal's approach to gain automatic pitch control by leading edge flaps lies in their variable deflection. Once the right tension of the rubber bands is set, the pitching moment curve potentially displays both: the relatively high pitching moments at low angles of attack with open flaps and a trim angle of attack in a useful interval around $\alpha = 10°$ to achieve the required lift with closed flaps. Measurements of the leading edge flaps hinge moments showed that the pretension of the rubber bands could be moderate to allow the automatic opening of the flaps and would still deflect the flaps sufficiently.

Wind tunnel and flight test data of the *Normal Soaring Apparatus*, the *Large Biplane* and the large *Experimental Monoplane* investigated here all show that Otto Lilienthal learned to design flying machines that allow stable flight within their individual flight envelope. The pitching moment data presented in this article demonstrated the potential of his invention for automatic pitch control and prove how well he understood the necessity to apply control surfaces to a monoplane with a wing span of nearly $9\ m$ in order to complement his type of weigh shift control. However, the stability, which was even higher than that of some later aircraft designs, was present only for flight attitudes that ensured largely attached flow on the wings and steady flight. Once the aircraft stalled, the stability vanished and the pilot had to react very rapidly by shifting his weight to the rising front and, when the stall occurred asymmetrically, to the rising wing's side. The reason for this limitation lies mainly in the design of the horizontal stabilizer and the tail. It was designed for flights in the vicinity of the ground and the flare landings that Lilienthal preferably performed. In his American patent description from 1895, which describes the world's first serial production aircraft the *Normal Soaring Apparatus*, he wrote: "*...on the latter is pivoted the tail in such a manner that it can freely turn upward, but finds downward a point of support on the fixed rudder. This mode of attaching the tail has the advantage that the tail will have no carrying action when the machine is employed like an ordinary parachute, thereby preventing from turning downward.*" This ability of the horizontal stabilizer to "*freely turn upwards*" and thereby to "*have no carrying action*", when the airflow acts from below, is wonderful when flying near the ground, but can become deadly when flying high. In case of a flare landing and stall that occurs near the ground, the result is a "pancake landing", a vertical fall, where the wings are leveled and act "*like an ordinary parachute*", as also commonly used for modern hang gliders. However, high flight altitudes require the aircraft to start diving in order to accelerate and recover. On August 9 in 1896, the day of his fatal crash at age 48, Otto Lilienthal flew his patented monoplane for the first time in several weeks, while he had concentrated on flying his *Large Biplane* in the meantime. At an altitude of approximately 15 m above ground, he was stopped by a wind gust and in spite of his experience didn't manage to lower the rising leading edge of his left wing by weight shifting as he had frequently done before. His deadly crash on that day confirmed the risk of flying at higher altitudes and outside the flight envelope at high angles of attack. His more than 2000 gliding flights had shown the stability and safety of gliding flights in calm air and near the ground.

2.3 Controllability

Considerable measurement noise was introduced by the long sting, which connected the 1:5-model to the measurement balance in the initial test setup in the SWG. This caused difficulties in isolating the aerodynamic effects of the control devices. Only the yaw moments due to the deflection of the rudder could be extracted reliably. Therefore, a second test setup in DLR's 1-m low-speed wind tunnel (1MG) was devised for the investigation of the wing warping and spoileron effects. The right half of the existing 1:5-model was mounted directly on a force- and moment balance and exposed to the flow through the wall of DLR's 1-m low-speed wind tunnel. Yaw and roll moments were recorded for a single flow velocity of $U_\infty = 8.5$ m/s. In addition to the measurement noise, the elasticity of the scaled model introduced unwanted aeroelastic effects. This was resolved by replacing the elastic Nylon lines, that braced the wing against the main frame, with steel wires. This more closely reproduced the elastic properties of the full-scale original, which proved to be essential for capturing the effect of wing warping and spoileron deflection.

The analysis of the glider's controllability focuses on the roll and yaw moment coefficients resulting from the deflection of the control elements. The control moments due to a deflection of the control elements were isolated by subtracting the results of the baseline configuration from the results of individually deflected control elements. It can be seen in Figs. 17 and 18 (red dashed lines) that adverse yaw occurs, when wing warping is applied. For the depicted wing warping, the leading edge was pulled inwards to reduce the wing's angle of incidence along its span, which decreases both lift and drag. The resulting roll and yaw moments act contrary to each other and have opposing signs in the investigated range of incidence angles from $6° \leq \alpha \leq 14°$.

The yaw moment coefficients for wing warping range from $c_n = -0.12$ to $c_n = -0.32$. The yaw moment coefficients of different rudder deflections are shown in Fig. 19. For the investigated set of deflections, a larger magnitude in yaw moment can be reached with the rudder up to an incidence angle of about 14°. It could therefore be used to compensate for the adverse

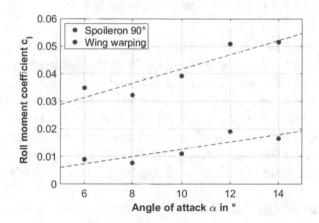

Fig. 17 Roll moment coefficients due to wing warping and spoileron deflection

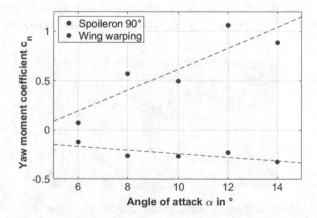

Fig. 18 Yaw moment coefficients due to wing warping and spoileron deflection

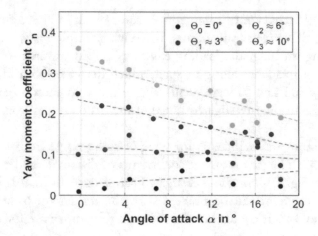

Fig. 19 Yaw moment coefficients due to rudder deflections

yaw of wing warping control actuation. The yaw moment curves depicted in Fig. 19 also show a decreasing effectiveness of the deflected rudder with an increasing angle of attack. This is caused by the increasing influence of the main wing wake at high angles of attack, which reduces the local flow velocity and therefore the side force generated by the deflected rudder. However, large rudder deflections $10° > \Theta > 6°$ create enough yaw moment to compensate the adverse yaw of the wing warping and should thereby allow for coordinated turns, when a correct combination of rudder deflection and wing warping is applied.

When the spoileron is actuated, the lift decreases while the drag is increased. As a result, roll and yaw moments with the same sign are created, as shown in Figs. 17 and 18 (blue dashed lines) for a spoileron deflection of 90° relative to the flight direction. This proverse yaw behavior, where the effects of roll and yaw act in the same turn direction, may nevertheless require the

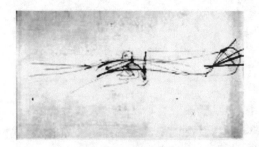

Fig. 20 Sketch for an elevator actuated by the pilot's lower back via lever and control wire [6]

additional use of rudder and possibly wing warping for properly coordinated turns, because the required ratio between roll and yaw depends on the turn rate. Otto Lilienthal sketched an actuated elevator the day before his fatal crash (see Fig. 20) but, to the author's knowledge, this variant was not flown by him.

3   Flight Tests

In 2021 tethered flights were performed at *Jockey's Ridge State Park* close to Kitty Hawk, North Carolina to gain practice and experience with the glider before the free flights were performed in California. The effect control surface actuation was limited for safety reasons, by the tethers attached to the wings. Although the wing tethers were frequently kept free of tension during the practice flights, they had to stabilize the glider in case of gusts. An additional towing line was used on several occasions to pull the glider forward, when the slope of terrain in wind direction was too shallow for pure foot launches (Figs. 21 and 22). Free downhill flight without any strings or ropes attached were performed by five test pilots at Marina Beachnorth of Monterey (CA).

Lilienthal stated that landing similar to turning the glider requires a counter-intuitive movement (Fig. 23). He reported during his lectures and

Fig. 21 Top: Otto Lilienthal near Berlin [10]. Below: Markus Raffel—in Nags Head near Kitty Hawk

Fig. 22 Practicing lateral and longitudinal control with limited control authority—tethered flight tests on the Outer Banks 2021

Fig. 23 Otto Lilienthal flying his *Experimental Monoplane* on the *Fliegeberg* 1895. *Photo* Neuhauss

in his written flight reports that the pilot has to move his legs backwards to pitch up and decelerate, even if his instinct advises him to position his feet forward when approaching the ground at higher speeds. However, the trim of the glider strongly influences this behavior. In the case of our test flights with the *Normal Soaring Apparatus*, and the *Large Biplane* it was sufficient to lean backwards with the upper body and therefore move the weight of the whole body to the rear.

In comparison to the *Normal Soaring Apparatus*, the rearward weight shift of the *Experimental Monoplane* had to be much more pronounced since the leading edge flaps kept were closed. During the tethered flights in the Outer Banks and later at Marina Beach in California the glider was tested only with firmly closed leading edge flaps as the correct tension of the rubber bands for opening at the right airspeed was unknown. From time to time this resulted in situations where the pilot had to shift his body weight drastically backwards (see Fig. 24). Otto Lilienthal quite obviously made similar experiences during his flights in 1895 when he experimented with the automatic pitch control system (see Fig. 25).

The use of a deflecting rudder was successfully demonstrated by Otto Lilienthal in 1895 (and again in 2021). Figure 26 shows the weight shift of Otto Lilienthal to the left and the rudder actuated by the hip cradle in the

Fig. 24 Michael Vaughn is turning the glider to flare by shifting his weight strongly backwards shortly before landing

Fig. 25 A flight where the angle of attack was low. Lilienthal is countering the dive by shifting his weight strongly backwards. *Photo* Neuhauss

Fig. 26 Practicing yaw control with the rudder actuated via the hip cradle—Otto Lilienthal, 1895

same direction. The same rudder deflection was initiated by the pilot during recent test flights. Wing warping could neither be tested during flight tests on the Outer Banks, nor during flight testing in California early 2022 (Fig. 27).

Fig. 27 Practicing yaw control with the rudder actuated via the hip cradle—Andrew Beem, 2022 (▶ https://doi.org/10.1007/000-6za)

4 Summary of Test Results

The scaled 1:5 model used during the wind tunnel tests had a wingspan of 1.76 m. It was shown by balance tests that a free stream velocity of 11.5 m/s was sufficient to lift the glider plus a person of Otto Lilienthal's weight ($\triangleq$80 kg) with a lift-to-drag ratio exceeding 5:1. The measured pitching moment coefficients, with a correction for the influence of the pilot's weight and location, proved that the longitudinal stability of this glider was adequate over a wider range of incidence angles. As long as the flow is attached the longitudinal flight stability is given. Differences between the measurements of default configurations and activated control surfaces showed that the wing warping generates adverse yaw, but a simultaneously activated ruder has the potential to counteract this effect.

Lilienthal tested many modern control methods on his 1895 experimental aircraft: wing warping for effective roll control, rudder deflection for yawing, and the control of both by means of unilaterally actuated spoilers (as used later as the primary control, for example, on the *Easy Riser* ultralight aircraft). However, Lilienthal's experiments may not have had a direct influence on the later development of powered flight, since they were neither combined, adjusted and tested to a satisfying level, nor widely published by him

5   Lilienthal's Progress in the View of the Wright Brothers

The combination of rudder deflection and wing warping is the control technique that the Wright brothers received a patent for in 1906 [13]. But since Lilienthal had described the method in detail only in his private correspondence with Alois Wolfmüller, but in the scientific publications just proposed turning the wings along their longitudinal axis, it must be assumed that the Wright brothers had developed the principle independently after reading the proposed literature of their days. In addition, they were the first to bring wing warping to success in a well-tuned combination with the rudder control and apply it to the mechanics of the biplane. So, since the Wrights were the pioneers who succeeded designing, building and flying an aircraft with three axis controls by a manually actuated elevator and wing warping and rudder actuated synchronously by a hip cradle in 1902, it might be interesting to recall what they thought about Lilienthal's contributions to flight.

In 1901 the Wright Brothers were frustrated after failing for the second year in a row to achieve the lift for their glider that their calculations predicted. In a speech on September 18 to the Western Society of Engineers, Wilbur suggested that *"the Lilienthal tables might themselves be somewhat in error."* He also questioned the accuracy of another value, namely the Smeaton coefficient as they used both the Lilienthal data and the Smeaton coefficient[2] in their formula for calculating lift. The Wrights would soon find that a Smeaton's coefficient value that was used at that time to derive the lift from the lift coefficient was too high. Lilienthal's data, the world's first systematically measured data of lift and drag coefficients of cambered wings, were quite accurate [14]. After presenting a paper to members of the western society and listening to comments from knowledgeable engineers, the Wright Brothers built a wind tunnel to develop their own aerodynamic data base, by systematically testing airfoils of widely different shapes and configurations, eventually also confirming Lilienthal's original data to a large extent. They incorrectly interpreted the Lilienthal tables by not understanding that the table only applied to one defined wing planform, and corrections for aspect ratio were only developed later. The wings that the Wrights used in 1900 and 1901 had aspect ratios between 3.5 and 3.3. Lilienthal used 6.8 for his test wing. Another misinterpretation of Lilienthal's data lay in the location of the point of maximum camber, which was close to the leading edge in the Wright Brothers wing design, whereas Lilienthal's wing was shaped like a circular arc.

---

[2] A more modern approach would use dynamic pressure, Prandtl's aspect ratio correction and Oswald's efficiency number instead of Smeaton's coefficient.

Writing her weekly letter to her father Milton, Katharine Wright also shared the latest scientific news in 1901 (see Fig. 28) [15]:

"*.. Will says he'll has to eat crow for, after all, he has discovered by accurate tests, that Lilienthal's tables are not far off.*" and some month later (in October 1901) Wilbur wrote to Chanute [16], "*It would appear that Lilienthal is very much nearer the truth than we have heretofore been disposed to think.*" Remembering Spratt's comments, the Wrights established new wind tunnel data for their glider in 1902, which helped them to develop their breakthrough flyer in 1903. Their wind tunnel data, the methodology and the increased understanding of all relevant parameters established the Wright Brothers as leading aeronautical engineers of their days as was Otto Lilienthal during his days.

The Wrights also wrote that, even though Lilienthal had made thousands of flights, they only added up to a few hours of training across a period of five years. To solve the problem of sustained flight, they needed a way to remain airborne for longer and therefore not only include an internal combustion engine that became more powerful over the years, but also enlarge the wing span and therefore to change from weight shift control, which was suitable for smaller aircraft, to an actuated elevator for pitch, wing warping for roll and rudder deflection for yaw control.

In 1912 Wilbur Wright concluded his assessment on Lilienthal's progress in this respect with the following sentences:

> "*Although he experimented for six successive years 1891–1896 with gliding machines, he was using at the end the same inadequate method of control with which he started. His rate of progress during these years makes it doubtful whether he would have achieved full success in the near future if his life had been spared, but whatever his limitations may have been, he was without question the greatest of the precursors, and the world owes to him a great debt.* "[17]

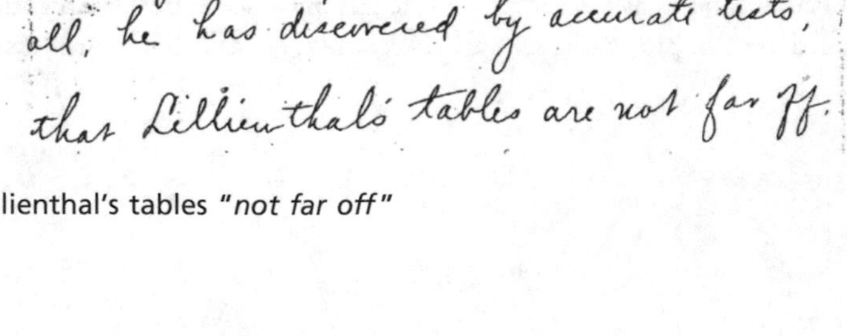
The boys are working very nights in their "scientific(?)" investigations. Poor Will says he'll have to eat crow for, after all, he has discovered by accurate tests, that Lillienthal's tables are not far off.

Fig. 28 Lilienthal's tables "*not far off*"

It is true that Otto Lilienthal had limitations and many other time-consuming activities like a family with four children, that he was the owner of a company with more than 30 employees, owned and operated a theater for members of his community. On the other hand, he designed, build and flew at least 12 different aircraft types in little more than five years and in contrast to the Wright brothers, who had flight reports, aerodynamic data, and Chanute's biplane wing design, which was flown already successfully years before they started, Lilienthal based his success mainly on his own experiments. When Lilienthal attempted to build his first powered aircraft in 1893, the only automobile that was produced in series, the Benz patent motor car *Velo* had 1.5 hp. In 1903, when the Wright brothers built their first powered aircraft, the automobile of that times, the Ford *Model A* already had five times the power (eight hp). It took the technical advancements of the preceding years, the talent of Charlie Taylor to build a 110 kg internal combustion motor with 12hp and some very advantageous weather conditions to fly in 1903. By 1908, the *Ford Model A* had 35hp.

However, the rapid progress in the development of internal combustion engines was made years after Lilienthal's flight experiments. Therefore, Lilienthal flew mostly without or in some limited cases with a very light weight self-designed two-cylinder carbon dioxide engine. At the end of his experiments with control methods, he stayed with the control method most adequate for hang gliders, which is still used by tens of thousands of pilots today: the weight shift control [18]. In October 1895 Lilienthal wrote to Wolfmüller: *"But in truth I am not convinced of these innovations—if the body is free to shift the center of gravity quickly enough, the same result can ultimately be achieved by other, simpler means. As always, practice remains key."* [19].

Acknowledgements The authors especially thank Simine Short and Billy Vaughn for their comments and support during the writing of the article and the founder of *Kitty Hawk Kites* (KHK), John Harris and the KHK experts Billy Vaughn and his son Michael, Dalton Burghalter, Larsen Christiansen, John Paul Gagnon and others who supported our flight tests. Markus Krebs' and Uwe Fey's help during the wind tunnel measurements and Anthony Gardner's CFD-computations are greatly appreciated. The weight shift controlled free flights in 2022 were generously supported by Barry Porter and JT Heineck, and conducted by the pilots Billy Vaughn, George Reeves, Jan Raffel (and Markus Raffel), while Andrew Beem flew with rudder deflection controlled via the hip cradle as well. His courage, skills and kindness are highly appreciated.

Bibliography

1. Anderson, J.D.: A History of Aerodynamics
2. Halle, G.: Otto Lilienthal: Flugforscher und Flugpraktiker, Ingenieur und Menschenfreund. 3rd edition, Düsseldorf (1976)
3. Halle, G.: Otto Lilienthal und seine Flugzeug-Konstruktionen. Munich (1962)
4. Heinzerling, W., Trischler, H.: Otto Lilienthal—Flugpionier Ingenieur Unternehmer. Munich (1991)
5. Lilienthal, O.: Der Vogelflug als Grundlage der Fliegekunst. Berlin (1889), reprinted 2003: ISBN 3-9809023-8-2.
6. Lilienthal, O.: Birdflight as the Basis of Aviation. New York (1911), reprinted 2001: ISBN 0-938716-58-1.
7. Lukasch, B. (ed.): Otto Lilienthal: Der Vogelflug als Grundlage der Fliegekunst. Berlin (2014)
8. Lukasch, B.: Otto Lilienthal auf Fotografien. Anklam (2016)
9. Nitsch, S.: Die Flugzeuge von Otto Lilienthal. Technik - Dokumentation – Rekonstruktion. Anklam (2016)
10. Runge, M., Lukasch, B.: Erfinderleben—Die Brüder Otto und Gustav Lilienthal. Berlin (2005)
11. Seifert, K.-D, Waßermann, M.: Otto Lilienthal - Leben und Werk. Hamburg (1992)
12. Schwipps, W.: Lilienthal: Die Biographie des ersten Fliegers. Munich (1986)
13. Schwipps, W.: Lilienthal: Der Mensch Fliegt. Koblenz (1988)
14. Schwipps, W.: Lilienthal und die Amerikaner. Munich (1985)

M. Raffel and B. Lukasch, *The Flying Man*, Springer Biographies,
https://doi.org/10.1007/978-3-030-95033-0

Appendix 3: References

1. Lilienthal, O., Birdflight as the Basis of Aviation, 1st edn. Longmans, Green, 1911, reprinted 2001: ISBN 0-938716-58-1. (Translation from German edition, Berlin 1889, "Der Vogelflug als Grundlage der Fliegekunst", reprinted 2003: ISBN 3-9809023-8-2.)
2. Chanute, O.: Progress in Flying Machines, (including the appendix "The Carrying Capacity of Arched Surfaces in Sailing Flight", translated from "Zeitschrift fur Luftschiffahrt und Physik der Atmosphäre", 1893), American Engineer and Railroad Journal, Courier Corporation (1894)
3. Wright, W.: Otto Lilienthal, Aero Club of America Bulletin, 1912 (published posthumously after Wilbur Wright died on 30 May 1912)
4. "Weight-shift Control Aircraft Flying Handbook" (FAA-H-8083–5) of the U.S. Departments of Transportation—Federal Aviation Administration, CreateSpace Independent Publishing Platform, 2013, ISBN 978-1490465319
5. Nitsch, S.: Die Flugzeuge von Otto Lilienthal, Otto-Lilienthal-Museum, Anklam, Germany (2016). ISBN 978-3-941681-88-0
6. Lilienthal, O.: US Patent for a "Flying Machine" No. 544,816, patented 20 Aug 1895, based on the German Patent for a "Flugapparat" No. 77916, patented 3 Sept 1893
7. Lilienthal, O.: "Brief an Wolfmüller", 3 Mar 1895, Deutsches Museum München HS1932-1, Otto-Lilienthal-Museum, Anklam, Archiv-ID 15904
8. McCroskey, W.J., Carr, L.W., McAlister, K.W.: Dynamic stall experiments on oscillating airfoils. AIAA J. **14**(1), 976, pp. 57–63
9. Kramer, M.: Increase in the maximum lift of an airplane wing due to a sudden increase in its effective angle of attack resulting from a gust. NACA-TM 678, translated from German from "Zeitschrift für Flugtechnik und Motorluftschifffahrt," vol. 23, no. 7 (1932)
10. Lukasch, B.: Otto Lilienthal auf Fotografien. Otto-Lilienthal-Museum, Anklam (2016). ISBN 978-3-941681-87-3

Appendix 4: References

1. Lilienthal, O.: US Patent Application for a "Flying Machine," Docket No. 544,816 (1895); also German Patent for a "Flugapparat," No. 77916 (1893)
2. Lilienthal, O.: Birdflight as the basis of aviation. Green and Co., Longmans (1911)
3. Culick, F.E.C.: Flight on the horizon: the pivotal year of 1896. AIAA J. **35**(2), 217–218 (1997). https://doi.org/10.2514/2.88
4. Crouch, T.D.: Octave Chanute and the Indiana glider trials of 1896. AIAA J. **35**(5), 769–775 (1997). https://doi.org/10.2514/2.7445
5. Chanute, O.: Progress in Flying Machines. Am. Eng. Railroad J. (1894)

6. Dees, P.: The 100-year Chanute glider replica, an adventure in education. In: 1997 World Aviation Congress, AIAA 975573 (1997). https://doi.org/10.2514/6.1997-5573
7. Culick, F.E.C.: The Wright Brothers: first aeronautical engineers and test pilots. J. Aircr. **41**(6), 985–1006 (2003). https://doi.org/10.2514/2.2046
8. Perkins, C.D.: Development of airplane stability and control technology. J. Aircr. **7**(4), 290–301 (1976). https://doi.org/10.2514/3.44167

Appendix 5: References

1. Jakab, P.L.: Otto Lilienthal: the greatest of the precursors. AIAA J. **35**(4), 601–607 (1997). https://doi.org/10.2514/2.154
2. Wienke, F., Raffel, M., Dillmann, A.: Wind tunnel testing of Otto Lilienthal's production aircraft from 1893. AIAA Aviation 2020 Forum, AIAA 2020-2738 (2020). https://doi.org/10.2514/6.2020-2738
3. Raffel, M., Wienke, F., Dillmann, A.: Flight-testing stability and controllability of Otto Lilienthal's monoplane design from 1893. J. Aircr. **56**(4), 1735–1742 (2019). https://doi.org/10.2514/1.C035399
4. Cherne, J., Culick, F.E.C., Zell, P.: The AIAA 1903 Wright 'Flyer' project prior to full-scale tests at NASA Ames Research Center. In: AIAA 38th Aerospace Sciences Meeting and Exhibit, AIAA 2000-0511 (2000). https://doi.org/10.2514/6.2000-511
5. Jex, H.R., Magdalano, R.E., Lee, D.: Virtual reality simulation of the '03 Wright Flyer using full scale test data. In: AIAA Modeling Simulation Technologies Conference, AIAA 2000-4088 (2000). https://doi.org/10.2514/6.2000-4088
6. Kochersberger, K., Ash, R., Britcher, C., Landman, D., Hyde, K.: Evaluation of the Wright 1901 glider using full-scale wind-tunnel data. J. Aircr. **40**(3), 417–424 (2003). https://doi.org/10.2514/2.3122
7. Kochersberger, K.B., Landman, D., Player, J.L., Hyde, K.W.: Evaluation of the Wright 1902 glider using full-scale wind-tunnel data. J. Aircr. **42**(3), 710–717 (2005). https://doi.org/10.2514/1.6955
8. Lawrence, B., Padfield, G.D.: Handling qualities analysis of the Wright Brothers' 1902 Glider. J. Aircr. **42**(1), 224–236 (2005). https://doi.org/10.2514/1.6091
9. Lawrence, B., Padfield, G.D.: Flight handling qualities of the Wright Brothers' 1905 Flyer 3. J. Aircr. **43**(5), 1307–1316 (2006). https://doi.org/10.2514/1.19607
10. Nitsch, S.: Die Flugzeuge von Otto Lilienthal, Otto-Lilienthal-Museum, Anklam, Germany (2016) (in German)
11. Schwipps, W.: Der Mensch fliegt. Bernard and Graefe (1988) (in German)

12. Savitzky, A., Golay, M.J.E.: Smoothing and differentiation of data by simplified least squares procedures. Anal. Chem. **36**(8), 1627–1639 (1964). https://doi.org/10.1021/ac60214a047

13. Lilienthal, O.: Über meine Flugversuche 1889–1896. VDI-Verlag, Ausgewählte Schriften (1996).(in German)

14. Weight-shift Control Aircraft Flying Handbook. U.S. Dept. of Transportation, Federal Aviation Administration Rept. FAA-H-8083-5 (2013). ISBN 978-1490465319

15. Kramer, M.: Increase in the maximum lift of an airplane wing due to a sudden increase in its effective angle of attack resulting from a gus. NACA-TM 678, 1932; also Zeitschrift für Flugtechnik und Motorluftschifffahrt, vol. 23, no. 7 (1932) (in German)

16. McCroskey, W.J., Carr, L.W., McAlister, K.W.: Dynamic stall experiments on oscillating airfoils. AIAA J. **14**(1), 57–63 (1976). https://doi.org/10.2514/3.61332

17. Kaufmann, K., Costes, M., Riches, F., Gardner, A.D., Le Pape, A.: Numerical investigation of three-dimensional static and dynamic stall on a finite wing. J. Am. Helicopter Soc. **60**(3), 032004.1–12 (2015). https://doi.org/10.4050/JAHS.60.032004

18. Perkins, C.D.: Development of airplane stability and control technology. J. Aircr. **7**(4), 290–301 (2012). https://doi.org/10.2514/3.44167

19. Lilienthal O.: Birdflight as the Basis of Aviation, 1st edn. Longmans, G. (1911) reprinted 2001: ISBN 0-938716-58-1. (Translation from German edition, Berlin 1889, Der Vogelflug als Grundlage der Fliegekunst, reprinted 2003: ISBN 3-9809023-8–2)

20. Lilienthal, O.: US Patent for a "Flying Machine" No. 544,816, patented 20 Aug 1895, based on the German Patent for a "*Flugapparat*" No. 77916, patented 3 Sept 1893

21. Lilienthal, O.: The Problem of Flying and Practical Experiments in Soaring. From the Smithsonian Report for 1893, Smithsonian Inst., Washington, DC, pp. 189–199 (1894)

22. Chanute, O.: Progress in Flying Machines (including the appendix "The Carrying Capacity of Arched Surfaces in Sailing Flight", translated from "Zeitschrift fur Luftschiffahrt und Physik der Atmosphäre", 1893). Am. Eng. Railroad J. Courier Corporation (1894)

23. Nitsch, S.: Die Flugzeuge von Otto Lilienthal, Otto-Lilienthal-Museum, Anklam, Germany (2016). ISBN 978-3-941681-88-0

24. Schwipps, W.: Lilienthal, Die Biographie Otto Lilienthal's, Deutsches Museum BN 46665

25. Raffel, M., Wienke, F., Dillmann, A.: Flying qualities of Otto Lilienthal's large biplane. J. Aircr. **58**(2), 413–419 (2019). https://doi.org/10.2514/1.C036022

26. Lukasch, B.: Otto Lilienthal auf Fotografien. Otto-Lilienthal-Museum, Anklam (2016). ISBN 978-3-941681-87-3

27. Lilienthal, O.: Brief an Wolfmüller, 3rd Mar 1895, Deutsches Museum München HS1932-1, Otto-Lilienthal-Museum, Anklam, Archiv-ID 15904

28. Oxford Aviation Academy: JAA ATPL 13: "Principles of Flight", Transair (2007)

29. Wienke, F., Raffel, M., Dillmann, A.: Wind-Tunnel Testing of Otto Lilienthal's Production Aircraft from 1893. AIAA J. **59**(4), 1342–1351 (2021). https://doi.org/10.2514/1.J059831

30. Wienke, F.: Aerodynamics of thin permeable wings (2020). ISBN 1434-8454, URL https://elib.dlr.de/138383

31. Wright, O., Wright, W.: Flying-Machine (1906). U.S. Patent No. 821393

32. Anderson, J.D.: A history of aerodynamics and its impact on flying machines, 8th edn. Cambridge University Press, Cambridge (2005). ISBN 0521669553

33. Wright, K., Wright, M.: 3, 25 Sept 1901, "The Wilbur and Orville Wright Papers" Family Correspondence, American Memory Collection, Library of Congress, Washington, DC (2003)

34. Wright, W., Chanute, O.: 12 May 1901, "The Wilbur and Orville Wright Papers" Octave Chanute Papers, American Memory Collection, Library of Congress, Washington, DC (2003)

35. Wright, W.: "Otto Lilienthal", Aero Club of America Bulletin (1912) (published posthumously after Wilbur Wright died on 30 May 1912

36. "Weight-shift Control Aircraft Flying Handbook" (FAA-H-8083-5) of the U.S. Departments of Transportation—Federal Aviation Administration, CreateSpace Independent Publishing Platform (2013). ISBN 978-1490465319

37. Schwipps, W.: Der Mensch fliegt - Otto Lilienthals Flugversuche in historischen Aufnahmen. Bernard und Graefe Verlag, Berlin (1988). ISBN 3-7637-5838-0

Printed in the United States
by Baker & Taylor Publisher Services